GEOLOGY AT ANU (1959–2009)

Fifty years of history and reminiscences

D. A. Brown Building 47, ANU

E PRESS

Published by ANU E Press
The Australian National University
Canberra ACT 0200, Australia
Email: anuepress@anu.edu.au
This title is also available online at: http://epress.anu.edu.au/geology_citation.html

---

National Library of Australia Cataloguing-in-Publication entry

Author: Rickard, M. J. (Michael John)

Title: Geology at ANU (1959-2009) [electronic resource] : 50 years of history and reminiscences / Mike Rickard.

ISBN: 9781921666674 (eBook : pdf) 9781921666667 (pbk.)

Subjects: Australian National University. Dept. of Geology--History
             Australian National University. Dept. of Geology--Anecdotes.

Dewey Number: 551.07119471

All rights reserved. No part of this publication may be reproduced, stored in a retrieval system or transmitted in any form or by any means, electronic, mechanical, photocopying or otherwise, without the prior permission of the publisher.

Book design and layout by Teresa Prowse, www.madebyfruitcup.com

Cover image: Weathered sandstone outcrop south of Darwin — painting by Glenys Eggleton

This edition © 2010 ANU E Press

GEOLOGY AT ANU (1959–2009)

Fifty years of history and reminiscences

Compiled by Mike Rickard
Edited by Judith Caton
Photographs by Richard Barwick

E PRESS

Contents

Plates	vii
Tables	xi
Abbreviations	xiii
Foreword	xv
Geology at The Australian National University, 1959–2009	xv
Editorial and Acknowledgments	xvii
The Canberra Setting	xix
Historical Highlights	xxi
Appreciation	xxiii
1. Introduction	1
2. The People	9
Teaching Staff	9
Technical and Administrative Staff	26
3. The Buildings	33
4. Teaching	41
Course Structure and Content	41
Student Fieldwork	50
5. Research	63
Staff Research	63
Research by Postdoctoral Fellows, Visiting Fellows and Research Assistants	87
Postgraduate Research	106
6. Major Research Projects	111
Geochemical Evolution and Metallurgy of Continents (GEMOC)	111
Australian Mining Industry Research Association (AMIRA)	113
Cooperative Research Centre for Landscape Evolution and Mineral Exploration (CRC-LEME)	115
Hot-Fractured-Rock Project	118
The Australian Marine Quaternary Program	120

7.	**Administrative Work, Honours and Publicity**	**125**
	Faculty Service	125
	Professional Bodies	126
	Editorial Work	131
	Honours	132
	Publicity	137
8.	**Student Activities**	**141**
	Student Awards	141
	Social Activities	143
	Field Excursions	152
	Graduate Destinations	154
9.	**The New Millennium**	**161**
	Course Structure	162
	Current Research	164
	Building and Equipment Support	166
	The Departmental Review of 2007	167
10.	**The Reunion and the Future**	**169**
11.	**Alumni News**	**173**
Appendix 1		**245**
	Visiting Fellows	245
Appendix 2		**251**
	List of BSc Honours Theses	251
Appendix 3		**265**
	List of Graduate Diploma Theses	265
Appendix 4		**267**
	List of MSc and MPhil Theses	267
Appendix 5		**271**
	List of PhD Theses	271

Plates

Cover: Weathered sandstone outcrop south of Darwin
Frontpiece: D. A. Brown Building 47, ANU

1   Science Faculty in 1960.
2   Tectonic Globe, Geology Department Foyer
3   William Smith Map, Geology Department Foyer
4   The Musical Buildings Cartoon
2.1   David Brown
2.2   Bruce Chappell
2.3   Ken Campbell
2.4   Allan White
2.5   Keith Crook
2.6   Ken Williams
2.7   Mike Rickard
2.8   Eric Conybeare
2.9   Tony Eggleton
2.10 John McDonald
2.11 Neil Williams
2.12 John Walshe
2.13 Warrington Cameron
2.14 David Ellis
2.15 John Tipper
2.16 Patrick De Deckker
2.17 Richard Arculus
2.18 Brad Opdyke
2.19 Leah Moore
2.20 John Mavrogenes
2.21 Stephen Cox
2.22 David Tilley
2.23 Prame Chopra
2.24 Jonathon Clarke
2.25 'Bear' McPhail
2.26 Ian Roach
2.27 Daniella Rubatto
2.28 Ken McQueen

2.29 Michael Elwood
2.30 Sara Beavis
2.31 Andrew Christy
2.32 Brian Kennett
2.33 Brian Harrold
2.34 Maree Coldrick
2.35 Ross Cliff
2.36 Henry Zapasnik
2.37 George Halford
2.38 Gill Lea
2.39 Zbigniew (Jack) Wasik
2.40 Tony Phimphisane
2.41 Jack Pennington
2.42 Ross Freeman
2.43 Tim Munson
2.44 Robin Westcott and John Vickers
2.45 Ulrike Troitzsch
2.46 Nigel Craddy
3.1   Campus map showing buildings occupied by the Geology Department
3.2   The Old Geology Building on University Avenue
3.3   Museum showcases in the foyer of the Old Building
3.4   Firemen visit the Old Building
3.5   The new wing of the Old Geology Building under construction
3.6   The flume in the Old Building
3.7   The New Geology Building: the D. A. Brown Building
3.8   The tearoom in the New Building on the occasion of Professor Campbell's seventieth birthday party
4.1   Manduramah field excursion, 1968
4.2   First-year excursion, 2003
4.3   David Ellis poses in front of the famous unconformity at Myrtle Beach
4.4   Students mapping at Bermagui
4.5   1968: Neil Williams leading his group underground at Broken Hill
4.6   Relaxing after fieldwork: students at the Silverton Pub, Broken Hill, ca 1994
4.7   The Golden Volcano, Tilba Tilba
4.8   Map of field camp locations in New South Wales
4.9   Allan White 'fishing' in Tuross Lake
4.10 Field camp at Wee Jasper, 2005
5.1   International Symposium of Vertebrate Palaeontology at the ANU, 1983
5.2   Ken Campbell and Dick Barwick on a fossil-collecting trip at Wee Jasper

5.3	Professor Chappell (centre) after the award of his DSc in 1990
5.4	Bruce Chappell modelling the latest in makeshift wet-weather gear to study granite
5.5	XRF analyser with modern automatic sample changer
5.6	A geologist at work
5.7	Mike Rickard 'examining' undeformed sediments at Merimbula
5.8	Ken Williams on fieldwork in western Tasmania
5.9	The GEROC logo designed by Dick Barwick for the group of retired palaeontologists working in the Geology Department
5.10	Picnic at Wee Jasper while on a fossil hunt in 1971
5.11	Fossil lungfish (*Barwickia downunda*)
5.12	Judith Caton and subject at Perth Zoo
5.13	*Stone Flowers*: paintings of early fossil angiosperms by Liz Truswell
5.14	Èva Papp in Antarctica
6.1	Cover of GEMOC second report
6.2	CRC-LEME poster
6.3	First-year students examine weathered granite at Island Bend in the Snowy Mountains, New South Wales
6.4	The Hot-Fractured-Rock Project explained
6.5	Lunch at Darling Harbour sponsored by CSIRO Marine Laboratories following the visit of the *RV Southern Surveyor* as part of the Geol 3018 Marine Geology Course held in August 2004 in Sydney
6.6	*RV Franklin* at Macquarie Wharf, Hobart, 1992
6.7	Emptying the box corer on the rear deck of *RV Franklin*
6.8	Map showing the location of many of the cores studied by members of the group
7.1	Dr Mike Rickard as University Marshall
7.2	The D. A. Brown Medal
7.3	Cover of the 1991 *Annual Report*
7.4	Cover of the 1992 *Annual Report*
8.1	Honours year 1987 outside the Old Geology Building
8.2	1993 Geology honours graduates with the ANU Marshall
8.3	Student Geological Society newsletter, September 1984
8.4	Wanted dead or alive: Tipper Dr J.C.
8.5	Examples of student T-shirt logos
8.6	Keith Crook in his field hat
8.7	ANU Australian Rules team, Perth Intervarsity, 1964
8.8	David Brown and Allan White in party mood
8.9	Cross-dressing: Warwick Crowe and Greg Miles
8.10	Honours Class of 1969 reunion plaque in the Geology Department foyer

Geology at ANU (1959–2009)

8.11   Mt Tarawera, New Zealand: end-of-year field excursion, c. 1992
9.1   Bachelor of Global and Ocean Sciences
9.2   The Geology Department's largest honours class—in 2001
10.1   Heads of department past and present
10.2   Early students and staff at the reunion dinner, University House, ANU
11.1   The Grogans in 1974
11.2   The Grogans in 2008
11.3   As President of the Peugeot Association of Canberra, Brad Pillans sometimes poses as Napoleon at the car club's 'Battle of Waterloo'
11.4   Richard Lesh, Simon Beams and Dave Jenkins, Directors of Terra Search Proprietary Limited

Back piece: End-of-year field excursion to Chile, January 2008: Jorisques Volcano with Licancabur Volcano behind on the left

Tables

2.1   Geology Department teaching staff
2.2   Technical and administrative staff in approximate date order
4.1   The 1990 course structure serves as an example of the Geology Degree course and the subject offerings for several years
5.1   Research assistants
5.2   National Research Fellows and Postdoctoral Fellows
7.1   Editorial work undertaken by staff members of the Geology Department
7.2   Honours conferred on members of the Geology Department
7.3   Recipients of the Stillwell Medal
7.4   Recipients of the A. B. Edwards Medal
8.1   Prizes awarded for outstanding achievement by undergraduate Geology students
8.2   Geology student prizes
    A. First-year geology
    B. Second-year geology
    C. Third-year prizes
        1. Anthony Seelaf Prize for fieldwork
        2. Irene Crespin Prize
        3. The Geophysics Prize
    D. The Ampol Prize for a student entering honours year, and the Western Mining (WMC) and Conzinc Riotinto (CRA) awards for honours fieldwork
8.3   The K. S. W. Campbell Prize
8.4   University awards
    A. ANU Medals awarded for exceptional achievement at honours level
    B. The Priscilla-Fairfield-Bok Prize for excellent achievement in science by a female student
8.5   The Robert Hill Memorial Prize
9.1   Geology courses offered in 2006–07
10.1 Attendees at the Geology Reunion Dinner
11.1 Contents of Alumni News in date order of graduation

Abbreviations

Actew	Australian Capital Territory Electricity and Water
AGSO	Australian Geological Survey Organisation
AIMS	Australian Institute of Marine Science
AJES	Australian Journal of Earth Sciences
AMIRA	Australian Mineral Industry Research Association
ANSTO	Australian Nuclear Science and Technology Organisation
ANU	Australian National University
Anutech	Australian National University business arm
ARC	Australian Research Council
ARGC	Australian Research Grants Committee
ANZAAS	Australia and New Zealand Association for the Advancement of Science
AusIMM	Australasian Institute of Mining and Metallurgy
ASIO	Australian Security Intelligence Organisation
ASX	Australian Stock Exchange
BMR	Bureau of Mineral Resources
BOZO	Department of Botany and Zoology
CCAE	Canberra College of Advanced Education
CRC	Cooperative Research Centre
CRES	Centre for Resource and Environmental Studies
CSIRO	Commonwealth Scientific and Industrial Research Organisation
DEMS	Department of Earth and Marine Science (ex Geology Department)
DETYA	Department of Education, Training and Youth Affairs
EEZ	exclusive economic zone
GA	Geoscience Australia
GEMOC	Geochemical Evolution and Metallogeny of Continents
GSA	Geological Society of Australia
LA-ICP-MS	laser ablation induced coupled plasma mass spectrophotometer
INAA	induced neutron-activation analyser
IGCP	International Geological Correlation Program
JCUNQ	James Cook University of North Queensland
LEME	Landscape Evolution and Mineral Exploration
MIM	Mount Isa Mines
MTEC	Minerals Tertiary Education Council

PESA	Petroleum Exploration Society of Australia
RSBS	Research School of Biological Sciences
RSES	Research School of Earth Sciences
RSPacS	Research School of Pacific Studies
RSPhysS	Research School of Physical Sciences
SHRIMP	sensitive high-resolution ion microprobe
SLEADS	salt lakes, evaporites and aeolian deposits
SREM	School of Resource and Environmental Management
SRES	School of Resource, Environment and Society
UC	University of Canberra
USGS	United States Geological Survey
WMC	Western Mining Corporation
XRD	X-ray diffraction analyser
XRF	X-ray fluorescence analyser

Foreword

Geology at The Australian National University, 1959–2009

This history of the Department of Geology at The Australian National University, which later became the Department of Earth and Marine Science, provides a fascinating insight into the development of the discipline and of the people involved over a 50-year period. The activity of the department is represented in the classroom and in the field, through the people, the research, student activities and the accomplishments of alumni.

Dr Mike Rickard is to be thanked for assembling the multifaceted contributions to this history, and for organising a celebratory dinner for the 50-year anniversary in March 2009, which many alumni managed to attend. Professor D. A. Brown, the Foundation Professor of Geology at the ANU, was present and delivered a memorable speech. Sadly, he died before the year was out.

The fiftieth anniversary of geology at the ANU came just after a major transition, at the beginning of 2008, with incorporation of the Department of Earth and Marine Science into an enlarged Research School of Earth Sciences, which was overseen by my predecessor, Professor Brian Kennett. In the new school, we build on the foundation of the past 50 years and seek to maintain and enhance the ANU's enviable reputation for strength in earth science research and education well into the future.

Professor Andrew P. Roberts
Director, Research School of Earth Sciences

Editorial and Acknowledgments

This work is as much reminiscence as it is history, with most of the information reported here gleaned from annual reports, newsletters or my memory. While the staff members deserve full biographies of their own, here, the contributions from many individuals, including staff and visiting fellows, provide brief information about their research and teaching.

Surprisingly, it was difficult to find hardcopies of all Faculty of Science *Annual Reports*. Maree Coldrick helped with access to departmental files and assisted with computing and student information. Thanks are also due to Maggie Shapley at the Menzies Archives, who kindly provided indexed early faculty *Annual Reports*; Angela Firth and Erica Walls-Nichols helped compile the list of student prizes; and Judith Shelley undertook the tedious task of checking the departmental lists of theses against the official lists in the Menzies Library. Professor David Brown provided an early history of the department, from 1959 to 1984. A major effort was made to contact former students, and brief reviews of their career activity since graduation make up a separate section of this history.

Christine Keller-Smith of the ANU Alumni Office kindly advertised the project and sought contributions from former students, as did Ron Hackney at Geoscience Australia, and Colin Simpson for the Australasian Institute of Mining and Metallurgy, while Sue Fletcher for the Geological Society of Australia arranged publicity to attract student contributions. David Brown, Ken Campbell, Tony Eggleton, Patrick De Deckker, 'Bear' McPhail and Robin Westcott kindly read and commented on sections of the report. 'Bear' McPhail, Patrick De Deckker and Peter Hancock provided information on the new millennium (Chapter 9); Ted Lilley contributed an account of geophysics teaching.

Photos and illustrations are acknowledged where possible. Several found in the department are of unknown origin. Surprisingly, there are no formal photographs of staff or students. George Halford provided a copy of Lithenea and other diagrams. Neil Williams, Ken Williams, Tony Eggleton, Patrick De Deckker, Ian Lambert, Sue Jephcott, Gavin Young, Andrew Lawrence, David Moore, Michael Conan-Davies, Anne Felton and Greg Miles provided photos of student activities; space precluded using several other submissions.

Richard Barwick has made a major contribution in printing and organising most of the illustrations for the manuscript. Judith Caton kindly undertook the onerous task of editing and, without her dedicated effort, this manuscript would never have been completed. Tony Eggleton, Anne Felton, Judith Shelley and Ted Lilley assisted with the final copyediting. An outside referee and the staff of ANU E Press are also thanked for their assistance. Mike Avent and Marilee Farrer (RSES) assisted in organising the reunion dinner held on 7 March 2009 at University House.

Mike Rickard, 2009

The Canberra Setting

The Commonwealth of Australia was formed by the federation of the Australian states in 1901. Canberra was founded as the national capital in 1913, and the Commonwealth Parliament moved to Canberra from Melbourne in 1927. With that move came Canberra University College, as part of the University of Melbourne. The Geology Department of this book was founded as part of Canberra University College.

In 1946, the Commonwealth Government founded The Australian National University as an institute of advanced studies to initially comprise four research schools. One of these, the Research School of Physical Sciences (RSPhysS), had a department of geophysics, to which in 1952 John C. Jaeger was appointed Foundation Professor. This department some years later expanded to a Department of Geophysics and Geochemistry and, in 1973, it became a research school in its own right: the Research School of Earth Sciences (RSES).

Canberra University College became part of the ANU—the School of General Studies—in 1960. Elsewhere in Canberra at that time professional geological activity grew with the Bureau of Mineral Resources (BMR), Geology and Geophysics, which was also established in 1946. The BMR geophysics branch, initially in Melbourne, moved to Canberra in 1966. In the activity described in this book, professional contacts with geologists and geophysicists from RSPhysS/RSES and BMR (later the Australian Geological Survey Organisation/AGSO; then Geoscience Australia/GA) were frequent and valued. In particular, several BMR personnel undertook higher degrees at the ANU, and many graduates of the Geology Department found challenging and satisfying careers with BMR/AGSO/GA.

Historical Highlights

1959	Professor D. A. Brown appointed
1963	Move to newly constructed Geology Building
1965	Extension to Geology Building
1969–70	Semester system initiated at the ANU
1983	K. S. W. Campbell appointed Professor
1980–83; 1988–93	M. J. Rickard Head of Department
1978; 1990; 1998	Departmental reviews
1989–94	Australian Mineral Industry Research Association (AMIRA) grants
1990	Association with School of Resource and Environmental Management (SREM)
1991	Move to Botany Building, the 'New Building'
1992	Anutech Geochemical Laboratory opened
1993–95	Joint field mapping camps with SREM and the University of Canberra (UC)
1994	R. J. Arculus appointed Professor
1995	First Landscape Evolution and Mineral Exploration (LEME) grant
1996	Geochemical Evolution and Metallogeny of Continents (GEMOC) grant
1997	Anutech Geochemical Laboratory closed
1998–2007	Professors D. J. Ellis, P. De Deckker and Dr D. C. McPhail Heads of Department
2004	Name changed to Department of Earth and Marine Science
2007	Departmental review
2008	Incorporated into RSES

Geology at ANU (1959–2009)

Major internationally acclaimed research achievements for the department have been

- the geochemistry of granites
- Pacific volcanics
- geochemistry of economic minerals
- the evolution of fishes over more than 400 my
- micropalaeontology and marine sedimentology that contribute to global climate-change studies
- the development of the weathered surface regolith
- laboratory and field studies on metamorphism
- detailed understanding of the processes of faulting.

Appreciation

A valedictory address by Tony Eggleton, on the occasion of the retirement of the Founding Professor, David A. Brown, in 1981

> Look then at what you have built, stone on stone,
> From nothing to high repute, known
> Even beyond great circles you have navigated
> Through students who have graduated.
> Not content with a rock to begin your creation
> You took a whole Cliff as the solid foundation.
> He's held it up firm, and it's never gone slack.
> The bones of the building came out of his back.
> Geologists must have the proper type-section
> Miss Healy first handled that job to perfection.
> Then books must be shelved, and the journals collected
> You might say it was double Val-ue[1] selected.
> To demonstrate science needs great dedication
> You added a Chappell for your congregation
> And asked him to analyse one bit of granite,
> It seems he won't stop 'til he's done the whole planet.
> At first Brown, D. A., had to do all the teaching
> And then Brown et Al.[2] gave a change to the preaching.
> If adding some colour was what you expected,
> With White you got purple as words interjected.
> Now wisdom, ye ken, might be found in a Scot
> Along with philosophy, science, the lot.
> But someone from Queensland? A dangerous gamble
> Which paid off as usual, so you got Campbell.
> The next, started off into sedimentation
> And gradually drifted to tectonisation.
> He studies orogenies (when the Earth shook)
> It's what you'd expect from a Quaker, Keith Crook.
> To teach about structure needs one you can trust
> To quickly distinguish a fold from a thrust

---

1   Val Herbst, librarian and draughtswoman.
2   Alan White.

Geology at ANU (1959–2009)

When cleavage and bedding are at the right spacing
Mike Rickard can tell which way up he is facing.
Next was Ken Williams, an epicurean
Who's tasted of sulfides from Stanford to Zeehan.
You struck the good oil when you found Eric Conybeare
Still one more space, so you added a Tony there.
Rocks that have value deserve to be fondled.
None loves them better than J. A. McDonald.
Buchites and boninites both need a hammerin'
That's why you brought over Warrington Cameron.
One who will make a new graptolite synthesis
Came with a beard, but we've since found a chin for Chris.
Some call him a tutor, some call him slave
Teaching nine courses, still grinning, is Dave.[3]
Now Alan and Bruce had an X-ray machine
Which had to be run, to be fed, and kept clean
And Jack got the job, and his work was quite stunning
Except that he thought he should do all the running!
The photos from X-rays are made without light,
(Astronomers take all their pictures at night)
But photos of fossils are taken con brio
So next from Otago you imported Leo.
It seems in a workshop each job's labelled 'RUSH',
And those in the queue should be given the brush,
A toolmaker needs to be part referee,
You turned up a master, the crafter Gil Lea.
But what matters more than machines so infernal?
Why, Min Mag, and Palaeontology Journal.
You cared for the library, kept it well funded
And saw that no centralist pillaged or plundered.
As libraries grow, all their books try to hide
To find them, you need an experienced guide
And guidance you got from the Rhodesian whiz
Mrs Jones (with her bike, and her violets, and Liz).[4]
Australia's Geology needed uniting,
You had all the staff, it was time to start writing
And poor Mrs Oliver typed the whole book

---

3   Demonstrator Dave Feary.
4   Mrs Jones was a Girl Guide Commissioner who had a picture of the Queen on her office wall.

Appreciation

That everyone knows as Brown, Campbell, and Crook.
But who'd do the diagrams, who's so intrepid?
Crafts girl extraordinary, young Judy Shepherd.
Some have quite newly learned Geologese
Like Radi, and Liliane. Judy, Louise,
And top of her class, Mrs Webber nee Rees
Who'll answer your questions on rocks, goats, or geese.
Minding machines needs a well-balanced team-man,
None does it better than rugby-fan Freeman.
And waiting for jobs gives you no cause for sorrow,
Joe did today what you'll ask on the Morrow.
To manage the chem lab you'll no need to fossick,
From over the creek you struck gold with Jack Wasik.
The slides that we get are so good they're phantasmic
When sectioned by Radi and Henry Zapasnik.
Minding Departments needs great elasticity
Sureness of touch without egocentricity
Being a Prof. is a piece of simplicity
When there's a girl at the front like Felicity.
Now the collections were growing colossal
With lots of nice minerals, even a fossil.
Specimens came from all over the place:
Australia, Westphalia, Sparta and Thrace.
From Greenland and Russia, from London and Salford,
Collected, curated, and labelled by Halford.
But many who came have since drifted away
Keith Ellis, Keith Massey, Paul Willis, Bob Grey,
Wendy and Sheralee, Barbara, and Anne,
Tall Mrs Drury and sweet Marianne.
Pam ran your office with great circumspection
While Greg drafted lines in her general direction.
How did the continents fit in Pangaea?
Belbin found out, he had UNIVAC's ear.
And don't we all miss that most loyal retainer
Who urned our affection with tea, Mrs Rayner?
And then Mrs Whyte, also quite indispensable
Even though usually incomprehensible.[5]
All of these people, and all of these words,

---

5   Mrs Whyte spoke broad Scots.

Geology at ANU (1959–2009)

> The rocks and the fossils, weren't just for the birds
> But rather for students, who came for your lectures
> And left full of knowledge, or sometimes conjectures.
> *[There followed a list of all students to that date]*
> A Department is more than the fame some go after
> It's care and variety, learning and laughter,
> At ANU 'Brown' means a rich polychrome
> All those who passed through called geology 'home'.

1
Introduction

In 1958, Canberra University College, which was then an academic adjunct of the University of Melbourne, decided to introduce a Faculty of Science with Departments of Botany, Chemistry, Geology, Physics and Zoology (and subsequently, Psychology and Geography, which had formerly been sited in the Faculty of Arts). To this end, five professors were appointed (Plate 1), and Dr David Brown, Reader in Geology at the University of Otago in Dunedin, New Zealand, arrived in Canberra for the beginning of the 1959 academic year. By 1960, the Prime Minister, Sir Robert Menzies, had become convinced that The Australian National University (an institute of advanced studies comprising research schools only), should include a School of General (Undergraduate) Studies, to wit the former Canberra University College with its new Faculty of Science. This merger is detailed in *The Making of the Australian National University 1946–1996* by S. G. Foster and Margaret M. Varghese (Allen & Unwin).

Plate 1 Science faculty in 1960. Front row centre (from left to right): the five original science professors, D. A. Brown (Geology), L. D. Pryor (Botany), A. N. Hambly (Chemistry), D. F. N. Dunbar (Physics), D. Smyth (Zoology). Back row: Geology technical staff Ross Cliff and George Halford (fourth and ninth from left), and students Ian Lambert and Fred Doutch (second and sixth from left).
Photographer R. E. Barwick (inset).

Geology at ANU (1959-2009)

At the beginning, teaching was undertaken by the Chemistry, Geology and Physics departments in the huts formerly occupied by the John Curtin Medical School near the Menzies Library, and, when the new buildings for Chemistry and Physics were opened in 1961, Geology moved temporarily into the Physics building.

A new purpose-designed building on the corner of North Road and the old University Avenue was occupied in 1963, and its eastern wing was added soon after (1965). It featured a main lecture theatre and separate laboratories for each year, which allowed students to have a home base for study. A Geology Branch Library—so essential for this scientific discipline—was also established but with continuing battles with the Central Library. We received generous contributions from the US Geological Survey, the British Museum (Natural History) and from several institutions in the Soviet Union. As the Research School of Earth Sciences (RSES) also built up its collection, a joint committee was set up to deal with location and duplications—mostly on an amicable basis.

Professor Brown attempted to build subject coverage with two staff in each sub-discipline. By 1968, a core complement of staff was in place, with two each in palaeontology, petrology and economic/petroleum geology, and one each in sedimentology, structural geology and mineralogy. The geophysics position was never filled, and this important field was covered by staff from RSES and the Bureau of Mineral Resources (BMR). Further development was forestalled by slow increases in student numbers, and later by competition from the newly set up Canberra College of Advanced Education (CCAE).

The wide experience of this staff nucleus welded into a friendly, competent team with a strong emphasis on teaching field geology, for which the local area was abundantly endowed. Unlike the northern hemisphere, here, the clash of teaching with the field season encouraged staff to undertake laboratory research. Nevertheless, many students were supervised in field projects in the outback with logistical support from the BMR and the Australian Geological Survey Organisation (AGSO) and exploration companies.

Collegiality was maintained with weekly seminars, Friday-night drinks at the old Staff Club and the practice of having the librarian circulate journals among all the staff. Inevitably, staff moved on—first to the much sought after field of economic geology as the mineral boom exploded, and later to promotional appointments elsewhere. Retirements also resulted in changes to the team. Emphasis in the courses offered also changed, with strong developments in geochemistry, computer applications, and regolith and marine studies.

Introduction

There have been three distinct administrative periods in the university: autocratic, democratic and bureaucratic. Initially, financial control was in the hands of a professorial board and, with a benevolent professor, the department flourished. With the aid of ANU and external grants, important major equipment was acquired, including X-ray fluorescence analyser (XRF), X-ray diffraction analyser (XRD), induced neutron-activation analyser (INAA), a large flume, rock saws, acid and palynological laboratories. End-of-year spending of remaining allocated funds enabled the purchase of a large globe, on which several vacation students painted ocean and continental tectonics from the then recently published international maps (Plate 2). Professor Brown acquired an original (fifth edition) *William Smith Geological Map of England, Wales and Lowland Scotland*—'The map that changed the world' (from the book by Simon Winchester, published in 2001) (Plate 3). This map and the globe became major display and teaching tools.

Plate 2 Tectonic Globe, Geology Department Foyer

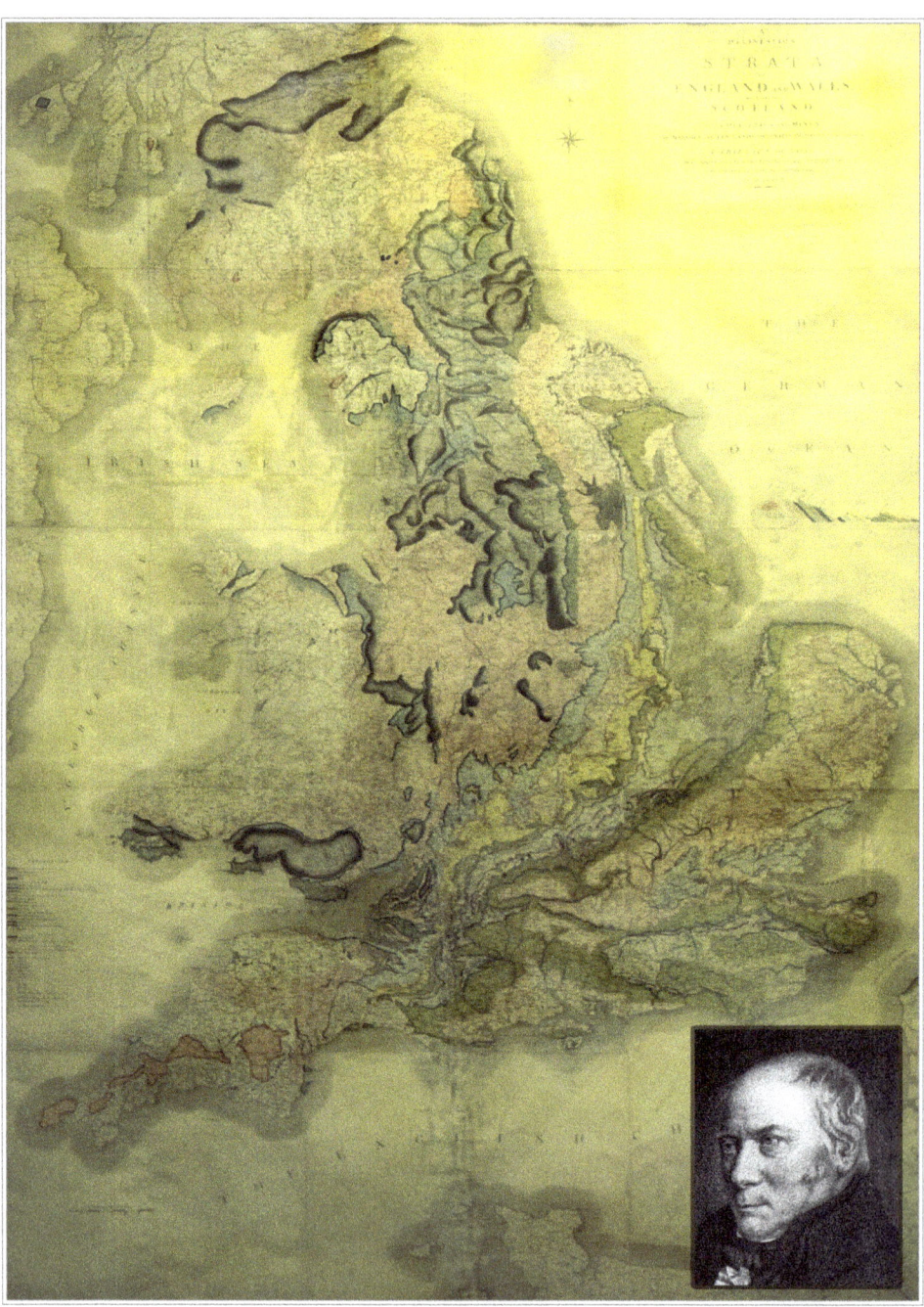

Plate 3 William Smith Map, Geology Department Foyer. 'The map that changed the world': William Smith's Geological Map of England, Wales and Lower Scotland (1815). Inset: Portrait of William Smith

Introduction

Initially, practical teaching specimens were obtained from the University of Melbourne, courtesy of Professor Sherbon Hills. Gifts, as well as purchases from the Kyancutta Museum in South Australia, augmented the teaching collections of minerals, rocks and fossils. The initial specimens for vertebrate palaeontology were plastic dinosaurs from cereal packets. Professor Brown also purchased a series of classic photographs of geological features from the UK Geological Survey. These adorned the first-year lab for many years, and some are still displayed in the present building.

During the democratic period, students gained access to most university committees, including finance. Students also demanded a say in course content and assessment—thus, faculty education committees were set up. A major change in course-work offerings came with the introduction of a semester system in 1969–70 (two 15-week semesters, instead of three terms) with a 20-point degree. This impacted on the large amount of fieldwork in geology courses as each field trip had to be allocated points or be incorporated into specific units.

Professor Brown relinquished the chair at the end of 1976. By then, university rules had changed and non-professorial heads of departments were allowed on three-year appointments. Ken Campbell served as head from 1977 to 1980. Then Mike Rickard was appointed head, from 1980 to 1982. Ken Campbell also served as Deputy Dean for one year, and then Dean of the Faculty of Science for two years, until he was appointed Professor in 1983. Ken stepped down as Head of Department in 1988; then Mike Rickard was reappointed, serving until Professor Richard Arculus joined the department in 1994. From 1997, several other staff took turns as Head of Department: David Ellis (1998–2002), Patrick De Deckker (2002–05), Richard Arculus (2005–06) and 'Bear' McPhail (2007).

From 1980 onwards, university finance was reduced and departmental budgets were cut severely. Research funding was granted through an Australian Research Grants Committee (ARGC) instead of departmental allocations and these funds had to make up for reduced departmental technical support. Over a few years, the department lost most of its support staff. The tea lady, typists, photographer, toolmakers, draughtswoman, museum curator, librarian and teaching demonstrators were phased out over the years.

Bureaucratic controls tightened over this period. The Department of Employment, Education and Training demanded detailed accounts of research activity; staff were required to submit biannual reports and to prepare teaching portfolios if seeking promotion. Eventually, elected deans were replaced by appointed ones—some from outside the university. The university set up an excellent Research Grants Office to assist staff in the increasingly important task of obtaining finance, but their instructions were often longer than the limit for grant applications.

In 1989, the ANU and the University of Canberra developed a joint Centre for Australian Regolith Studies—then the only cooperation between the two universities. Although small, it paved the way for further cooperation. Next, under the Dawkins Review, came a move for the ANU to merge with the University of Canberra, and we spent much time designing possible course structures. We also began joint teaching in three subjects (structural geology, sedimentology, and economic geology) and later joint field camps, also with ANU Geography, at Boorowa, New South Wales (see Chapter 4). The proposed merger eventually failed and, several years later, the UC geology courses were reduced and eventually closed in 2003. There is currently a memorandum of understanding with the UC on cooperation for a Capital Water Program.

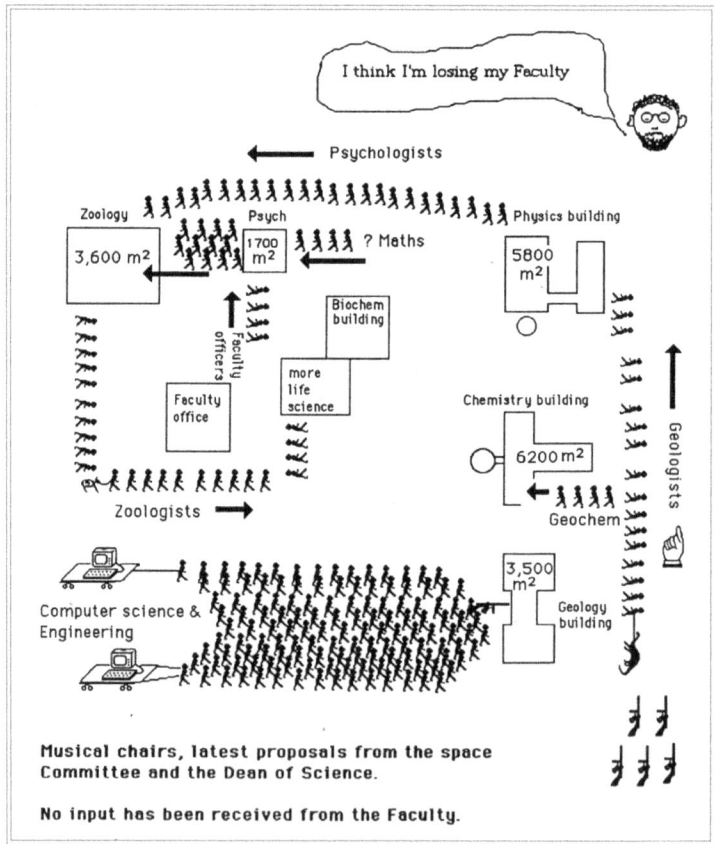

Plate 4. The Musical Buildings Cartoon. Tony Eggleton's response to the proposed ousting and move of the Geology Department to the Botany Building.

Introduction

Then came a major disruption. It was decided in 1991 to move Geology out of its original building to allow housing of the new Engineering Department, and to consolidate 'science precincts'. Other departments also suffered in this major upheaval; Botany moved in with Zoology and some Life Sciences with Biochemistry; the Psychology building became a science teaching block; and Geology—avoiding being split into chemistry and geography—reluctantly moved into the old Botany building late in 1991 (Plate 4 and Plate 3.1).

Almost a year was spent designing new laboratories and offices, and housing the museum collections. A major blow was the loss of the library, in spite of a faculty committee recommendation that it be kept. The collection was split—part going to RSES and the rest to the Central Science Library. The map collection was sent to AGSO and a new computer lab was established in the room allocated for the library. After much disruption to research, the department eventually settled in. The central courtyard became a social focus, and, on opening day in March 1992, Professor Brown planted a symbolic new 'fossil' maiden-hair tree (*Ginkgo biloba*) to replace the ones that had been near the original huts and the old Geology building (see Plate 7.4).

In the meantime, Geology was encouraged to associate with Geography and Forestry in a newly formed School of Resource and Environmental Management (SREM)—fields in which we had little expertise. The first-year course in geology had already been redesigned to incorporate a semester of physical geography. Remote sensing was also offered—first in Geography, then later in Geology.

Three major research programs were awarded to the department. In 1995, the Centre for Australian Regolith Studies (ANU/UC) together with AGSO and the CSIRO Division of Exploration and Mining were awarded a Commonwealth Government Cooperative Research Centre for Landscape Evolution and Mineral Exploration (CRC-LEME). Also in 1996, an Australian Research Council (ARC) National Key Centre for the Geochemical Evolution and Metallogeny of Continents (GEMOC) was granted to Macquarie University's and the ANU's Geology departments in collaboration with CSIRO Exploration and Mining and AGSO. The mining industry also supported the department's research in petrology and economic geology with several industry (Australian Mineral Industry Research Association) grants over several years. Ampol, Conzinc Riotinto of Australia (CRA), and Western Mining have supported honours students and, in 1984, Esso funded 50 per cent of a tutorial post.

A gradual increase in Postdoctoral Fellows and Visiting Fellows—especially after AGSO made many staff redundant—boosted the department's research output to one of the best in the faculties. Success in obtaining research grants and ship-time has been outstanding, and publication citation indices for several staff are among the highest in Australia.

Staff retirements have prompted a variety of functions. A large gathering at Bruce Hall farewelled Professor Brown. Tony Eggleton as MC recited his own poetic valediction, in which he mentioned all our old students (see 'Appreciation'). The retirements of Ken Campbell and Keith Crook were marked with special national seminars on palaeontology and sedimentology respectively, and, for Mike Rickard, a dinner in the Union was graced by a poetic slide show—by Tony again—based on the hymn *Jerusalem*: 'and did those feet in ancient times walk upon England's mountains green', and so on. Tony Eggleton's own retirement dinner was held at Vivaldi Restaurant on campus. Other farewell lunches were held in University House. A large gathering (92) of granite workers, in Canberra for the second Hutton Symposium in 1991, met for dinner at Rydges Lakeside in tribute to Bruce Chappell's granite research. Several American guests were bewildered by the pavlova dessert, with its passionfruit and kiwifruit toppings.

Staff retirements and new appointments—especially the appointment of Professor Arculus in 1996—ushered in a change of focus towards volcanic geochemistry and marine geology. Then, in 2004, the department's name was changed to the Department of Earth and Marine Sciences, and subsequently a first-year course in marine science was offered.

In 1978 and 1990, the department was subjected to major reviews, which it passed with flying colours; importantly, the continuance of the chair was recommended. Another review was conducted in 2007 that again commended the department for its research and teaching efforts, but expressed concern at its budgetary situation and recommended a merger with RSES to form a School of Earth Sciences (see Chapter 9).

2
The People

Teaching Staff

Plate 2.1 David Brown
Plate 2.2 Bruce Chappell
Plate 2.3 Ken Campbell

David Brown

David Brown (Plate 2.1) graduated MSc in 1937, after his appointment as Field Geologist to the New Zealand Geological Survey in 1936. He transferred to NZ Petroleum Exploration in 1938. Following the outbreak of World War II, he served with the New Zealand Expeditionary Force in New Zealand, then in 1941 transferred to the Royal Navy Fleet Air Arm for service in the United Kingdom as an observer during the Norwegian Campaign. At the end of the war in 1945, he took a honourable discharge and was awarded a postgraduate scholarship at the Imperial College of Science and Technology and the British Museum (Natural History), London, graduating (PhD, Diploma Imperial College London) in 1948. He then returned to the New Zealand Geological Survey until he was appointed Senior Lecturer in Geology at Otago University, Dunedin, in 1950. In 1958, he was appointed Professor of Geology at Canberra University College (ANU). David Brown retired and was appointed Emeritus Professor at the beginning of 1981. David Brown died in November 2009 at the age of ninety-three.

Bruce Chappell

Bruce Chappell (Plate 2.2) came from the University of New England in 1960, where he was working on local granites. In Canberra, he set up a major geochemical laboratory and carried out analyses to complete his PhD as a staff candidate. Bruce Chappell subsequently carried out major geochemical research on the granites of south-eastern Australia. He was awarded a DSc from the ANU in 1990, was promoted to a personal chair in 1992, and was made a Fellow of the Academy of Science in 1998. He retired in 1997.

Ken Campbell

Ken Campbell (Plate 2.3) graduated from the University of Queensland. He then worked for a year with the Geological Survey and the University of Queensland on the 40-mile map of the State. In 1951, Ken was a mathematics teacher at Albury Grammar School for a year. He then moved to the University of New England, where he taught palaeontology and stratigraphy until 1961, when he joined the ANU. He initially specialised in brachiopods, but later worked on trilobites and then Devonian fish. In 1983, he was appointed Professor of the department and also elected a Fellow of the Australian Academy of Science. Professor Campbell retired in 1992, but continues his work in the department as an Emeritus Professor and Visiting Fellow to this day.

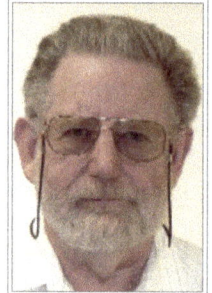

Plate 2.4 Allan White
Plate 2.5 Keith Crook
Plate 2.6 Ken Williams

Allan White

Allan White (Plate 2.4) graduated from Adelaide University in 1951. He was one of Sir Douglas Mawson's last honours students, mapping granites in the eastern Mount Lofty Ranges. He took a PhD from King's College, London, supervised by Professor Wally Pitcher, before becoming a lecturer at Otago University, New Zealand, in 1957. He was appointed as Senior Lecturer at the ANU in 1960, joining the small original team in the department. Together with Bruce Chappell, he taught petrology and made major field and geochemical studies of granites in south-eastern Australia. Allan resigned in 1971 to take up the new Chair of Geology at La Trobe University in Victoria. In 1996, Allan returned from retirement to become the third director of the Victorian Institute of Earth and Planetary Sciences—the combined association of the Melbourne universities' Earth Science departments for teaching and research. Allan died in September 2009.

Keith Crook

Keith Crook (Plate 2.5) graduated from Sydney University with a BSc (1954) and MSc (1956), then did a PhD at the University of New England (1956–59), followed by postdoctoral studies in Melbourne (1959) and the University of Alberta, Canada (1959–61), before coming to Canberra in 1961 to teach sedimentology and stratigraphy. He also introduced a study of soils and later marine geology into the course. Keith Crook undertook research in New Guinea and Tumut (New South Wales) followed by marine work in the Manus Basin. This research was concerned with the tectonic development of sedimentary basins. The department was unsuccessful in a 1987 bid to set up a special research centre in marine geoscience, so, frustrated by not being able to extend his program, Keith took early retirement in mid-1992. Following this, Keith was appointed to a position at the University of Hawai'i as Science Program Director of the Hawai'i Undersea Laboratory. He returned to the ANU as a Visiting Fellow in mid-2004, where he continues his research.

Ken Williams

Ken Williams (Plate 2.6) graduated from the University of New England before working with CSIRO Minerals Division in Melbourne. He came to Canberra in 1961 to teach economic geology and mineralogy and to continue his research on the mineral deposits of western Tasmania. Ken Williams was awarded his PhD by the ANU in 1968. He left the department in 1969 to become an Associate Professor at Stanford University, USA. He returned to Sydney University in 1974 and, since 1991, has been a private consultant.

Geology at ANU (1959–2009)

Plate 2.7 Mike Rickard
Plate 2.8 Eric Conybeare
Plate 2.9 Tony Eggleton

Mike Rickard

Mike Rickard (Plate 2.7) graduated from Imperial College London in 1954. He carried out field structural studies in Donegal for a PhD and in the Canadian Appalachians as a Postdoctoral Fellow at McGill University. After working for the Geological Survey of Fiji, he came to Canberra in early 1963 to teach structural geology and tectonics, and to assist with field geology. Mike served as Head of Department in 1980–82 and again in 1988–93. He retired at the end of 1997.

Eric Conybeare

Eric Conybeare (Plate 2.8) was born in the United Kingdom, but became a Canadian citizen, graduating from the University of Alberta; then, after war service, he took a PhD from Washington State University. He worked in Ghana and then for the Shell Oil Company in western Canada. Eric Conybeare came to Canberra to join the Petroleum Branch of the BMR, but soon relocated to the ANU in 1964 to teach petroleum geology and basin analysis. Eric donated the petroleum drill bit that is still used as a doorstop outside the lecture theatre. He developed interests in conservation issues and transferred to the Centre for Resource and Environmental Studies (CRES) in 1978, where he wrote two books. He died shortly after his retirement in 1982.

Tony Eggleton

Tony Eggleton (Plate 2.9) took a PhD from Wisconsin University, USA, after graduating from the University of Adelaide. He was appointed in 1966 to teach mineralogy and he also organised petrology field camps. In the 1970s, Tony Eggleton extended his studies to weathering and played a leading role in a major research investigation of the regolith. He was awarded a DSc from Adelaide University and promoted to Professor in 1999. Together with Professor Graham Taylor, he wrote *Regolith Geology and Geomorphology*, a book published in 2001. He retired that year but continues his association with the department as a Visiting Fellow.

The People

Plate 2.10 John McDonald
Plate 2.11 Neil Williams
Plate 2.12 John Walshe

John McDonald

John McDonald (Plate 2.10) was appointed to replace Ken Williams to lecture in economic geology. He graduated in Canada from the University of Manitoba, and took a PhD at the University of Wisconsin. In 1971, John McDonald transferred from the CSIRO's Baas Becking Lab to the ANU. He resigned in 1980, returning to Canada to work as a consultant, where he is currently involved in diamond exploration.

Neil Williams

Neil Williams (Plate 2.11) graduated from the ANU in 1969 and took a PhD at Yale University in the United States. He was seconded from RSES for the second semester 1980 to teach the economic geology courses at the ANU and CCAE. Neil was popular with students as he rewarded their efforts with prizes of chocolate frogs. He joined industry in 1981 for several years before returning to government service in Canberra. He is currently CEO of Geoscience Australia (GA).

John Walshe

John Walshe (Plate 2.12) was appointed in 1981 to replace Neil Williams, who had temporarily taken over from John McDonald. He graduated with a PhD from the University of Tasmania in 1977; his doctoral dissertation was on the geochemistry of the Mount Lyell copper deposits in western Tasmania. He subsequently worked as a research associate with Dr Mike Solomon. They co-authored a paper, 'The formation of massive sulfide deposits on the sea floor', that anticipated the discovery of black smokers. As a lecturer in economic geology at the ANU, his work with graduate students culminated in a special volume on the Metallogeny of the *Tasman Fold Belt System of Eastern Australia*, published in *Economic Geology* in 1995. He resigned in 1995 to join CSIRO.

Geology at ANU (1959-2009)

Plate 2.13 Warrington Cameron
Plate 2.14 David Ellis
Plate 2.15 John Tipper

Warrington Cameron

Warrington Cameron (Plate 2.13) graduated from Cambridge University and was appointed an Associate Lecturer (Tutor) after the resignation of Allan White. This was a temporary position with limited tenure. After five years of teaching petrology, he transferred to the museum curator's position and continued assisting with practical classes in petrology and field excursions for several years until leaving the department in 1988.

David Ellis

David Ellis (Plate 2.14) graduated from the University of Melbourne and completed his PhD at the University of Tasmania. He replaced Warrington Cameron as Associate Lecturer in Petrology. Later, the position was re-established and made permanent. David teaches metamorphic petrology and has set up a laboratory for experimental work, and also runs an annual petrology field trip through Victoria and out to Broken Hill in New South Wales. He was promoted to a personal chair in 2000.

John Tipper

John Tipper (Plate 2.15) was appointed to replace Eric Conybeare. He graduated from the University of Cambridge (BA, 1970) and the University of Edinburgh (PhD, 1974); he then held a Royal Society European Programme Postdoctoral Fellowship at the University of Bonn (1973–74). He moved to the University of Kansas to work as a mathematical geologist at the Kansas Geological Survey (1974–78), then to University College Galway to teach palaeontology and stratigraphy (1978–83). John Tipper came to the ANU in 1983 to lecture in sedimentary-basin analysis, as well as participating in stratigraphy, sedimentology, petroleum and coal geology courses and on field camps—most memorably at Taemas and at Rangari. He moved to Germany (University of Freiburg) in 1992 to take up a position as Professor of Historical Geology, Palaeontology and Sedimentology.

The People

Plate 2.16 Patrick De Deckker
Plate 2.17 Richard Arculus
Plate 2.18 Brad Opdyke

Patrick De Deckker

Patrick De Deckker (Plate 2.16) was appointed to teach palaeontology after Professor Campbell retired in 1988. He graduated from Macquarie and Adelaide universities. His speciality is micropalaeontology applied to climate change, with major studies on lake-bed and marine faunas—the latter involving extensive ocean cruises. Patrick De Deckker was promoted to a personal professorship and was awarded a DSc by the University of Adelaide in 2002, and an Australia Medal in 2006.

Richard Arculus

Richard Arculus (Plate 2.17) was appointed Professor in 1994 after the retirement of Ken Campbell. He graduated from Durham University, UK, with First-Class Honours in 1970, and received his PhD in 1973. He worked at the University of Michigan, and served as Professor at the University of New England before coming to the ANU. Richard specialises in volcanic geochemistry, particularly of the Pacific Ocean region.

Brad Opdyke

Brad Opdyke (Plate 2.18) was appointed in 1994 to replace Keith Crook to lecture in sedimentology. He graduated from the University of Michigan. Brad Opdyke specialises in the study of carbonates and has started a research project on the reef islands off northern Australia. He has also taken over the running of geology field camps.

Geology at ANU (1959–2009)

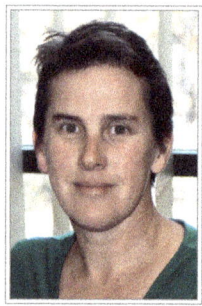  

Plate 2.19 Leah Moore
Plate 2.20 John Mavrogenes
Plate 2.21 Stephen Cox

Leah Moore

Leah Moore (Plate 2.19) was appointed in 1990 as an Assistant Lecturer (Tutor) mainly to deal with first-year students—the first and, until recently, only female member of academic staff! She graduated from the University of Auckland and took her PhD at the ANU (1997) in Regolith Studies. In 1997, Leah was appointed First-Year Coordinator; she also assisted with second-year sedimentology and petrology classes. She greatly enthused students and led two field excursions to New Zealand. She also attempted to recruit students to her sport of 'underwater hockey'. She was awarded the first K. S. W. Campbell Prize for excellence in teaching. From 1994 to 1997, Leah was a Research Fellow at Monash University, before returning to the ANU as a Visiting Fellow with the Regolith Group and Lecturer in the School of Resource, Environment and Heritage Science at the University of Canberra. Her subsequent career is outlined in Chapter 11.

John Mavrogenes

John Mavrogenes (Plate 2.20) graduated with a PhD from Virginia Polytechnic Institute and State University in 1994. He was a Western Mining Corporation-sponsored Postdoctoral Fellow at RSES with the Petrochemistry and Experimental Petrology Group before taking up a joint appointment with the Geology Department in 1997. He was promoted to Senior Lecturer in 2002 and Associate Professor in 2008. John specialises in the geochemistry of economic minerals, and he has taught the third-year Ore Deposits Course since 1995.

Stephen Cox

Stephen Cox (Plate 2.21) gained his BSc (Hons) from the University of Tasmania in 1973. He subsequently worked for two years as a structural geology specialist with the Geological Survey of Tasmania, and then undertook a PhD at Monash University. From 1980 to 1984, he was a Postdoctoral Fellow at Monash University, with funding provided by a CSIRO Postdoctoral Fellowship and

The People

a Queen Elizabeth II Fellowship. In 1984, Stephen moved to RSES at the ANU as a Research Fellow, then Senior Research Fellow in the Petrophysics Group. In 1995, he took up the position of Professor of Geology at the University of Newcastle (New South Wales). In 1998, he returned to the ANU on a professorial appointment held jointly in the Department of Geology and RSES.

Plate 2.22 David Tilley
Plate 2.23 Prame Chopra
Plate 2.24 Jonathon Clarke

David Tilley

David Tilley (Plate 2.22) graduated from Flinders University (South Australia) and took his PhD at the ANU in 1996. He was appointed a Research Fellow in 1995; then in 1996, he became an Associate Lecturer to assist with mineralogy and the newly developing Regolith Studies. He resigned in 1998. David's subsequent career is outlined in Chapter 11.

Prame Chopra

Prame Chopra (Plate 2.23) completed his BSc at Newcastle University (New South Wales) in 1976, and then undertook his PhD research in Geophysics at RSES, graduating in 1980. Prame Chopra was a Postdoctoral Fellow at RSES in 1980–81, and then at Cornell University, New York, from 1982 to 1983, studying high-temperature rock physics. He was an ARC appointment to Geochemical Evolution and Metallogeny of Continents (GEMOC) to teach geological information systems and geophysics and was promoted to Reader in 1999. Prame worked with the Hot-Fractured Rock Project (see Chapter 6). He resigned in 2006 to participate in the commercialisation of that project.

Jonathon Clarke

Jonathon Clarke (Plate 2.24) received his BSc from the University of Tasmania, and his PhD from Flinders University (South Australia). He was appointed in 1999 as an Associate Lecturer (CRC-LEME) for three years. He assisted with lectures and labs in regolith, mineralogy and sedimentology, and also with first-year practical classes and field excursions.

Geology at ANU (1959–2009)

Plate 2.25 'Bear' McPhail
Plate 2.26 Ian Roach
Plate 2.27 Daniella Rubatto

Derry Campbell 'Bear' McPhail

Derry Campbell 'Bear' McPhail (Plate 2.25) graduated from the University of British Columbia with BSc in 1980 and MSc in 1985, followed by a PhD from Princeton University in 1991. From then until 2002, he was a Postdoctoral Fellow, Lecturer and Senior Lecturer at Monash University, where his research focused on hydrothermal geochemistry and environmental geoscience, and he developed environmental science courses and degrees before coming to the ANU in 2002 as a Reader. Bear served as an executive and key researcher for the Cooperative Research Centre for Landscape Environments and Mineral Exploration (CRC-LEME). In the department, he has led activities related to CRC-LEME, in addition to developing and teaching courses in regolith, natural resources, groundwater and environmental chemistry. At present, his studies are centred on groundwater research for mineral exploration and environmental geoscience. In 2007, Bear served as Head of Department.

Ian Roach

Ian Roach (Plate 2.26) took his BSc and PhD at the University of Canberra studying the Monaro basalts. In 1997, he became a Research Assistant with CRC-LEME, and, from 2001 to 2002, he held a joint lecturer's appointment at the ANU and UC. From 2003, he was appointed as a Mineral Council of Australia lecturer in regolith geology. He left the department in 2007.

Daniella Rubatto

Daniella Rubatto (Plate 2.27) graduated from the University of Turin (Italy) in 1994, and took a PhD from the Swiss Federal Institute of Technology (*Eidgenössische Technische Hochschule*/ETH), Zurich, in 1998. She was a Postdoctoral Fellow at RSES until joining the Department of Earth and Marine Science (DEMS) in 2002. Daniella has taught geochemistry in the department since 2007.

The People

Plate 2.28 Ken McQueen
Plate 2.29 Michael Elwood
Plate 2.30 Sara Beavis

Ken G. McQueen

Ken G. McQueen (Plate 2.28) joined the department as Reader in Applied Geochemistry in February 2003. This was a fractional full-time position, which he held until January 2008. During this time, Ken McQueen was also Assistant Director of CRC-LEME. His main activities were research into new methods and approaches to mineral exploration in the Australian regolith, undergraduate teaching in regolith geology and supervision of honours and postgraduate students. He also ran a number of short courses in regolith geology and mineral exploration for the Minerals Tertiary Education Council (MTEC) through the ANU. Ken maintains an active connection with the department as an Adjunct Professor.

Michael Elwood

Michael Elwood (Plate 2.29), before his appointment to the ANU as Research Fellow/Lecturer in 2006, held the following positions: Teaching Assistant in the Chemistry Laboratory, University of Otago (New Zealand), from 1994 to 1997; Postdoctoral Fellow, University of Liverpool (Oceanography), from 1998 to 1999; Research Fellow, University of Canberra (Eco-Chemistry Lab), from 2000 to 2001; Postdoctoral Fellow at the National Institute of Water and Atmospheric Research (New Zealand), from 2001 to 2004; and Research Scientist at the same institute from 2004 to 2006.

Sara Beavis

Sara Beavis (Plate 2.30) graduated with her PhD from the University of NSW School of Mines, Department of Engineering Geology in 1992. She came to the ANU in 1995 to work with the Centre for Resource and Environmental Studies (CRES) and then the Fenner School. In 2007, she was seconded to DEMS to work with Landscape Evolution and Mineral Exploration (LEME) for two years. She teaches a third-year coastal environment earth sciences course and contributes to first-year lectures. Her research focuses on acid-sulfate soils and the physical characteristics of the interaction of water and the regolith.

Plate 2.31 Andrew Christy
Plate 2.32 Brian Kennett

Andrew G. Christy

Andrew G. Christy (Plate 2.31) graduated with a BA (1984) and PhD (1988) in Earth Sciences from Cambridge University. After two postdoctoral years at Cambridge, he moved in 1990 to the Chemistry Department at the University of Leicester, commuting to the synchrotron at Daresbury to study properties of materials at high pressure, and then worked full-time at the synchrotron from 1993 to 1994. He joined the ANU in 1994 as a Research Fellow with the Disordered Materials Group in the Research School of Chemistry, and worked there from 1994 to 1999. From then until 2001, he worked with the Applied Mathematics Department in the Research School of Physical Science and Engineering. In 2001, he became Research Officer in DEMS and, in 2009, a Lecturer (25 per cent fractional). During his time at the ANU, he has taught a first-year course in field geology, second-year courses in mineralogy, lithosphere and analytical chemistry, a honours course in solid-state chemistry, and a Masters course in the nuclear fuel cycle.

Brian Kennett

Brian Kennett (Plate 2.32) is the current Director and Professor of Seismology in the Research School of Earth Sciences. As director, he has executive control of the newly combined (2008) DEMS and RSES. He received his PhD in Theoretical Seismology from the University of Cambridge in 1973, and then spent 1974–75 as a Lindemann Fellow at the Institute of Geophysics and Interplanetary Physics, University of California at San Diego, and subsequently was a lecturer in the Department of Applied Mathematics and Theoretical Physics at the University of Cambridge. He moved to RSES in 1984 as Professorial Fellow. Professor Kennett was President of the International Association of Physics of the Earth and

Planetary Interiors (IASPEI) from 1999 to 2003. His research has covered a wide range of topics in seismology, from exploration work with seismic reflection to the free oscillations of the Earth, and he has developed Earth models that are now standard in many applications—for example, earthquake location. He has received recognition through the Adams Prize of the University of Cambridge (1983), as Fellow of the American Geophysical Union (1987), as Fellow of the Australian Academy of Sciences (1994), the Jaeger Medal of the Australian Academy of Sciences (2005), as Fellow of the Royal Society, London (2005), the Murchison Medal of the Geological Society of London (2006), the Beno Gutenberg Medal in Seismology from the European Geosciences Union (2007), and the Gold Medal for Geophysics from the Royal Astronomical Society (2008).

Geology at ANU (1959-2009)

Table 2.1 Geology Department teaching staff

Professors
D. A. Brown#, MSc, PhD, DIC (London), Hon. FGS Aust.; died 2009
K. S. W. Campbell#, FAA, MSc, PhD (Qld), Hon. FGS Aust.
R. J. Arculus, BSc, PhD (Durham)
B. W. Chappell#, FAA, MSc (NE), PhD (ANU), DSc (ANU), Hon. FGS Aust.
R. A. Eggleton#, BSc (Adelaide), PhD (Wisconsin), DSc (Adel.)
D. J. Ellis, MSc (Melbourne), PhD (Tasmania)
P. De Deckker, AM, BA, MSc (Macquarie), PhD (Adelaide), DSc (Adelaide)
B. L. N. Kennett, FRS, FAA, PhD (Cambridge)
S. Cox*, BSc (Tasmania), PhD (Monash)
Readers (now Associate Professors)
A. J. R. White#, BSc (Adelaide), PhD (London); died 2009
K. A. W. Crook#, MSc (Sydney), PhD (NE), BA (ANU), Hon. FGS Aust.
M. J. Rickard#, ARCS, BSc, DIC, PhD (London), Hon. FGS Aust.
P. N. Chopra#, BSc (Newcastle), PhD (ANU)
D. C. McPhail, BSc, MSc (UBC), PhD (Princeton)
J. Mavrogenes*, BS (Beloit), MS (UMR), PhD (Virginia Tech.)
K. McQueen, BSc (UNE), PhD (UWA)
Senior Lecturers
K. L. Williams#, BSc MSc (UNE), PhD (ANU)
E. Conybeare, PhD (Alberta); died 1982
J. McDonald#, BSc (Manitoba), PhD (Wisconsin)
J. L. Walshe#, BSc, PhD (Tasmania)
J. C. Tipper#, BA (Cambridge), PhD (Edinburgh)
B. N. Opdyke, AB (Columbia), MS, PhD (Michigan)
Lecturers
N. Williams#, BSc (ANU), PhD (Yale)
I. Roach, BSc, PhD (UC)
M. Ellwood, BSc (Otago), PhD (Otago)
Associate Lecturers
D. B. Tilley#, BSc (Flinders), PhD (ANU)
C. L. Moore#, BSc (Auckland), PhD (ANU)
W. E. Cameron#, MSc (Melbourne), PhD (Cambridge)
J. Clarke#, BSc (Tasmania), PhD (Flinders)
A. G. Christy, BA (Cambridge), PhD (Cambridge)
D. Rubatto, BSc (Turin), PhD (ETH)
S. Beavis, PhD (NSW)

retired or resigned
* joint appointment with RSES

Demonstrators

Demonstrators were part-time PhD students employed to also act as teaching assistants for laboratory work and fieldwork. Initially, they assisted with first-year, petrology and palaeontology courses, and later with structural geology and sedimentology courses, as well as other courses offered by the department.

E. G. (Peter) Wilson

E. G. (Peter) Wilson (1962), one of the first ANU MSc students, came to the department from the BMR. He assisted with general tutoring of first-year students.

Tony Taylor

Tony Taylor graduated from King's College, London, and spent time with a Greenland expedition before working on granites in Victoria from 1963. He was a great raconteur. Tony unfortunately contracted tuberculosis and left the department in 1968 to become a schoolteacher in Scotland.

Bob Day

Bob Day graduated from the University of Queensland and became a Senior Demonstrator in Palaeontology at the ANU in 1964. He completed his PhD in 1969 as a staff candidate. His subsequent career is outlined in Chapter 11.

Alex Grady

Alex Grady (see Plate 4.1: back row) graduated from Sydney University and did a PhD on Structural Geology at Otago University before coming to the ANU as a Senior Demonstrator from 1966 to 1968. He spent four years in exploration, and then taught at the new Flinders University, from where he carried out research in East Timor. Alex Grady and Ron Berry (now at the University of Tasmania) caught the last plane out of Dili as the Indonesian invasion threatened. Alex later became Head of School and Faculty; then, from 1997 to 2006, he was Professor and Head of Campus at the Central Queensland University, Bundaberg. He is now retired and living in Tasmania.

John Moss

John Moss graduated in 1963 from Reading University, UK, and worked at the ANU for his PhD on the detailed sedimentology of sand deposits (1968). He demonstrated in first-year geology classes and in sedimentology courses. John Moss, together with Keith Crook, was instrumental in setting up the flume (see Plate 3.6). He subsequently worked with the CSIRO Division of Soils, Canberra.

Karl Wolf

Karl Wolf was appointed in 1963 as a Senior Demonstrator in Palaeontology after completing his PhD at Sydney University. He subsequently (1964) went on to teach in the United States and Canada, and has carried out extensive consulting and editorial work.

Graham Taylor

Graham Taylor began at the ANU as a vacation scholar over the summer of 1965–66, working with Keith Crook. He was appointed Senior Demonstrator in 1969 and left in March 1977 to take up a post as Lecturer in Geology and Soils at CCAE. He was awarded his PhD from the ANU that same year. His thesis was on the Darling River and its sediments—what could be called a depositional model for suspended-load streams. He then spent 1973–74 mostly off work after a serious car accident whilst returning from a student field trip in Tasmania. Over the years, he became involved in many things, some of which involved the department; these included periods as a Fellow and, with Tony Eggleton, the beginning of the Centre of Australian Regolith Studies (CARS), which eventually led to the UC and the ANU going into CRC-LEME. Graham retired from the UC as a Professor Emeritus in 2004, but continues to work with Tony Eggleton on various regolith research topics.

Bob Gunthorpe

Bob Gunthorpe was awarded a PhD from the University of New England in 1970. He was a Senior Demonstrator in Petrology from 1969 to 1970, and later undertook postdoctoral research at the University of Western Australia. Bob carried out exploration work in Namibia, before returning to Canberra to work in the Parliamentary Library. He currently works as a consultant.

Hans Hensell

Hans Hensell was Senior Demonstrator in Petrology, and then became a consultant in the dimension-stone industry. He organised the tearoom's granite tables, clocks for sale by students, and the floor display in the foyer of the New Building.

John Funk

John Funk graduated from Missouri University and came to Canberra with an MSc from Imperial College London in Structural Geology (1969). He worked with the BMR mapping in central Australia and he assisted with structure classes for 10 years. He left to take up a schoolteaching post in Sydney.

Ross Both

Ross Both did his MSc research at the University of Tasmania before coming to the ANU as a Senior Demonstrator and PhD candidate in 1966. He assisted with first-year geology classes and mineral-deposit practicals, as well as giving some lectures while Ken Williams was on study leave. He left in 1968 to take up a lectureship in economic geology at the University of Adelaide. His PhD on the geochemistry of minor elements in the Broken Hill Lode was successfully completed in 1970.

Chris Jenkins

Chris Jenkins taught in the Geology Department from 1981 to 1982, helping Ken Campbell in his sabbatical years. He was very active in the department, running weekly seminars, called 'Geobabbles', and assisting with palaeontology practicals. He came to the ANU from Cambridge University (UK), fresh from completing a PhD on the graptolite stratigraphy of classic Murchison–Sedgwick areas in Wales. Subsequent to his time at the ANU, Jenkins taught at James Cook University, Townsville, and then returned to Sydney (where he had earlier taken his BSc Hons) and joined the fledgling Ocean Sciences Institute with Gordon Packham. The years following were adventurous and productive, with marine expeditions through Polynesia, to Antarctica, the Atlantic and Indian oceans, and the Baltic Sea—for geophysical mapping, sedimentology and deep-sea photography. By then, GIS technologies were emerging as the db SEABED project gradually took form. In 1999, it was adopted by the US Geological Survey (USGS) to map the US exclusive economic zone (EEZ). In 2002, he was offered a position at INSTAAR (Colorado University) in Boulder, USA, where db SEABED matured into an international collaborative research program. It is now the most extensive and detailed database of the global ocean-bottom substrates, and is used in a multitude of ways, from demersal shark ecology to adaptive underwater communications.

Ian Smith

Ian Smith graduated from the University of Auckland. He came to the ANU from the BMR in 1971 to study PNG volcanics for his PhD (awarded in 1976). He demonstrated in petrology courses and assisted with fieldwork, especially on Mount Dromedary, at Tilba Tilba. Ian's subsequent career is outlined in Chapter 11.

Gerry Reinson, Doug Mason, Dave Holloway, Brian Chatterton and Dave Feary

Gerry Reinson, Doug Mason, Dave Holloway, Brian Chatterton and Dave Feary also served for periods as demonstrators. The careers of the first three are outlined in Chapter 11.

By 1972, demonstrator positions were abolished and were replaced with fixed-term (five-year) Assistant Lectureships (Tutors), of which the department had three: Cameron, Tilley and Moore (see previous section: Teaching Staff). From this time on, the department had a large influx of Research Assistants and Postdoctoral Fellows, mostly supported by outside funds, but these people were not generally involved with teaching. From 1972, PhD and honours students worked as demonstrators, especially for first-year classes.

Technical and Administrative Staff

The nature of the teaching and research in geology demanded a wide range of technical assistance, in addition to the normal administrative support staff of departmental secretaries, including draughting, photography, museum curating, library, field services (including a vehicle fleet), lapidary technicians, as well as skilled assistants for X-ray and geochemical analyses. Newly designed apparatus also needed skilled toolmakers. Initially, there were 17 general staff members; then, following massive reductions, numbers fell from 12 in 1988 to nine in 2007. The departmental tea lady was first to go, without anyone realising that she also did all the photocopying work. This increased the workload of typists and other staff. One staff member came in on a Sunday to photocopy notes for a Monday lecture. On finding the machine reading 'out of fluid', he refilled it from the only visible can and completed his work. Come Monday morning, the office staff found the copier gummed up; it had been refilled with floor polish!

Plate 2.33 Brian Harrold (Research Assistant, IT, 1986 – present)
Plate 2.34 Maree Coldrick (Departmental Administrator, 1989 – present)

The rapid acquisition of personal computers made typists redundant; by 1995, there were none. From two small Apple Macs in 1990, all staff had computers within a few years. Photography, draughting and toolmaking were centralised in the faculty. Field camps had to be run with minimal technical support; one dean thought we should replace them with videos! The librarian went with the loss of the library in 1991, and the worst was the later loss of the museum curator, who had looked after the valuable teaching collections. Some technical geochemical positions were supported by outside grants and through Anutech, as was the invaluable new position of a computer expert (**Brian Harrold**; Plate 2.33).

The general staff have made many valuable contributions to the smooth running of the department. **Mrs Oliver**, one of the first departmental secretaries, typed many manuscripts, including helping with typing Professor Brown's translation work. Head Technical Officers (HTOs) **Ross Cliff** (Plate 2.35), **Derek Butterfield** and **Henry Zapasnik** (Plate 2.36) handled the departmental finances until this responsibility was passed to **Maree Coldrick** (Plate 2.34) in 1998. Maree is the longest-serving Departmental Administrator. In 2006, her duties were extended for a year to assist with CRC-LEME administration.

Geology at ANU (1959–2009)

Plate 2.35 Ross Cliff (HTO, 1959–87)
Plate 2.36 Henry Zapasnik (Technical Officer and HTO, 1980–98)

Ross Cliff set up the original technical support system for the department. **Derek Butterfield**, from the Chemistry Department, was involved with the formalisation of procedures for handling radioactive materials in the Geochemistry Lab. **Henry Zapasnik** was responsible for much of the design and movement logistics for the transfer to the Botany building. He also invented new techniques for processing micro-fossils, for which he was awarded one of the first ANU Honorary Master of Science degrees for technical support. Henry also initiated cooperation with the new National Aquarium, and organised a fishing expedition to collect lungfish in Queensland. **George Halford** (Plate 2.37) started as a Technical Officer (TO) with Geophysics (later RSES), and then transferred to the Geology Department. He completed a part-time BSc in 1966, followed by an MSc in 1971. He became a Fellow of the Gemmological Society of Australia in 1973. George was appointed Museum Curator in 1966, building up excellent international collections of teaching and research material for the department. He was also a skilled petrology demonstrator, and assisted with many field trips. In the 1960s mineral boom, George also became the unofficial adviser to an informal investment group in the department.

Plate 2.37 George Halford (Museum Curator and Research Officer, 1959–89)

The People

Plate 2.38 Gill Lea (TO Toolmaker, 1964–80)
Plate 2.39 Zbigniew (Jack) Wasik (TO Geochemistry, 1960–89)
Plate 2.40 Tony Phimphisane (TO, 1989 – present)

As toolmaker, **Gill Lea** (Plate 2.38) helped Bruce Chappell design and produce a prototype sample changer for the XRF machine, which was quickly commercialised, and later (1971) won a design award. This avoided the all-night duties for students carrying out rock analyses. **Joe Morrow** took over this role after Gill retired. **Zbigniew (Jack) Wasik** (Plate 2.39) quietly ran the Geochemistry Lab and carried out most of the analyses on water samples for more than 40 years. **Tony Phimphisane** (Plate 2.40) first patiently learnt the skills to run the geochemical support services, and then learnt the skills to take over the lapidary services after John Vickers retired. **Jack Pennington** OAM (Plate 2.41) ran the X-ray machines for 20 years, and, a veterans' athletic coach, became the departmental fitness adviser. **Ross Freeman** (Plate 2.42) also ran the X-ray machines for many years until he left to set up his own X-ray service company. From 1992, the Geochemistry Laboratory was run as a commercial operation by Anutech under the management of **Ulrick Senff**. It closed on Professor Chappell's retirement in 1997.

Plate 2.41 Jack Pennington (TO XRD, 1963–83)
Plate 2.42 Ross Freeman (TO XRF, 1971–84)
Plate 2.43 Tim Munson (Tutor and Museum Curator, 1985–2000)

Chris Foudoulis ran the XRD laboratory and took over the photography lab after Leo Seewin's retirement. Then, from 1990, he ran the Shrimp microprobe for AGSO/GA. **Robin Westcott** (Plate 2.44) started with the lapidary group and later became a skilled operator for the XRD laboratory. She also served as occupational health and safety officer, student and staff counsellor, and field-camp organiser; students fought for her vegetarian casseroles.

Geology at ANU (1959–2009)

Plate 2.44 *Robin Westcott (TO XRD, 1986–2002) and John Vickers (TO, 1988 – present)*

One of the early technicians, **Bob Gray**, was a veteran 'Rat of Tobruk', and a visiting Russian was convinced he was an Australian Security Intelligence Organisation (ASIO) agent because he had noticed him lighting his pipe with a match from a box bearing an ASIO badge! **John Vickers** (Plate 2.44) worked in the Lapidary Lab and as a field hand for outback work. He also gave instructions on driving field vehicles. **Craig Price**, although part-time, drove students to Gladstone in Queensland several times for the Heron Island excursions. **Radi Popovic** made many contributions to fieldwork, especially as an excellent long-distance driver and in laying out the survey grids at Taemas each year in preparation for the first-year mapping exercise—in spite of the thistles! **Elsie Jones** served us very well as librarian in spite of her African violet collection contaminating the Palynology Laboratory. Recently (from 1990 to the present), **Val Elder**, after working as a fossil preparator, has contributed much valuable voluntary service cataloguing the museum collections.

Tim Munson (Plate 2.43), **Henry Zapasnik** and **Tony Phimphisane** collected and made up a special plaque with Parliament House granite and Red Hill limestone that was presented by our Olympic team as an Australian contribution for a commemorative shrine in Seoul, Korea.

Of the current technical staff, **Ulli Troitzsch** (Plate 2.45) runs the XRD laboratory and **Linda McMorrow** the XRF operations. **Nigel Craddy** (Plate 2.46), together with **Andrew Christy** (Plate 2.31), provides general support for all technical operations in the department.

The People

Plate 2.45 Ulrike (Ulli) Troitzsch (TO XRD, 2003 – present)
Plate 2.46 Nigel Craddy (TO, 1988 – present)

Several of the technical staff were awarded study leave to enhance their expertise. Ross Cliff and Henry Zapasnik visited the United Kingdom and the United States to arrange purchase of new rock-sectioning machines; Ross Freeman visited geochemical laboratories in South Africa; Brian Harrold attended an international information technology (IT) conference in Europe; and John Vickers attended a national technology meeting in Albury, New South Wales.

Field vehicles played an important role in the department's activities. These were looked after by Radi Popovic, and then John Vickers. Initially, we ran Land Rovers and VW Kombi wagons. Professor Brian Mason, who visited annually for several years, lent us his long-wheel-base Land Rover emblazoned with 'Smithsonian Museum Meteorite Expedition' for use between his visits. Later, we settled on Mitsubishi four-wheel-drive Star wagons that seated eight, and Subarus for field supervision. Radi used his contacts among car dealers to get us good exchange bargains. Later, however, Central Purchasing took over this task and equipped us with rented brand-new four-wheel-drives to use for bush bashing! We, luckily, had very few accidents over the years in spite of heavy vehicle use; John Vickers can be thanked for good driver training in recent years. For a full list of the Geology Department's technical and administrative staff, see Table 2.2.

Cleaning staff are not listed although in the early days they were on the department's allocation. For several years, Joe Morrow's son, Bill, looked after the building.

Table 2.2 Technical and administrative staff in approximate date order

Departmental Secretaries	Draughting	Technical Officers: lapidary	Technical Officers: geochemistry and X-ray
Mrs J. Kelly-Healy	Mrs Val Herbst	Bob Gray	Nola Dechazel
Mrs J. McGinley	Judy Shepard/Davis	Sue Jephcott	Zbigniew (Jack) Wasik*
Mrs Oliver*	Mrs J. L. Sleep	Mrs De Smet	Maureen Kaye
Mrs D. Thrift	Greg Harper*	J. T. Boyd	Jack Pennington OAM*
Felicity Adams/Chivas*	Ms A. Northmore	Ms B. Cameron	Ross Freeman*
Pam Coote	Lilian Widmeier*	Ms D. Dillon	Chris Foudoulis*
Maree Coldrick*		Ms M. Whyte	Elizabeth Webber*
Judy Papps (LEME)	**Museum**	K. F. C. Ellis	Robin Westcott*
Deborah Bordeau (LEME)	George Halford*	D. Markovic	Roy Doyle*
Jenna Leonard (LEME)	Warrington Cameron	Keith Massey*	Tony Phimphisane*
Judith Shelley (LEME)	Tim Munson*	H. Zapasnik*	Berlinda Crowther
		Mrs S. B. Black	Dr Ulli Troitzsch
Typists	**Librarians**	Mrs C. M. Ridgeway	Dr Andrew Christy
Barbara Mitchell	Mrs Val Herbst	P. E. Willis	Linda McMorrow
Cherilee Bell	Mrs D. Pearce	P. A. Scott	G. Olley (Anutech)
Wendy E. Christian	Mrs Val Taylor	Radi Popovic*	D. Steele (Anutech)
Ros Graham-Brown	Mrs J. Attwood	H. D. Hensel	Dr Ulrich Senff (Anutech)
Anne Kilner	Mrs Elsie Jones*	N. C. Robertson	M. Marsh (Anutech)
M. I. Harden	Mrs Sirkka Spanari*	I. F. Atkinson	
Mrs B. E. Richards		J. N. Olley	**Workshop**
Mrs H. Drury*	**IT Support Services**	S. Bygrave	Gill Lea*
M. J. Metzler	Brian Harrold*	D. B. Collins	Joe Morrow*
F. L. McConchie		R. A. Creaser	
V. A. O'Neil	**Head Technical Officers**	Ms K. J. Gibbs	**Photography**
Elaine Byrne	Ross Cliff*	M. B. Mills	Leo Seewen*
Louise Smith	Derek Butterfield	Robin Westcott*	Chris Foudoulis*
Liz Lemon	Henry Zapasnik*	John Vickers*	
Mrs M. C. MacDougall	Norman Fraser*	Craig Price	**Tea ladies**
Mrs Mary Hope*	John Seeley	Norman Fraser*	Mrs Rayner
	Sarah O'Callaghan	J. Van Daele	Mrs Whyte
		C. Rumble	
		John Seeley	
		Nigel Craddy*	

* long-term staff

3
The Buildings

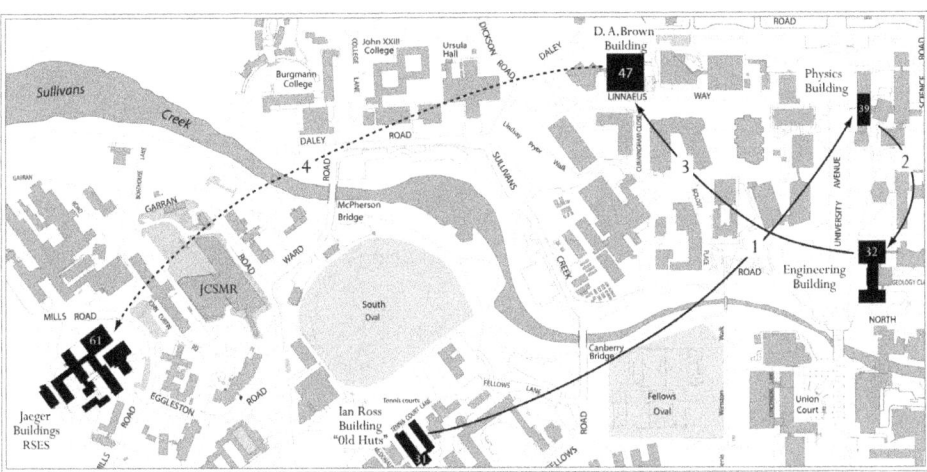

Plate 3.1 Campus map showing buildings occupied by the Geology Department. 'Old Huts', Building 31 (1959–60); Physics Building, Building 39 (1962–63); Old Geology Building, Building 32 (1963–91); D. A. Brown Building, Building 47 (1991 – present); RSES, Building 61, future location?

R. E. Barwick modified from original map courtesy of Paul Sjoberg, ANU Facilities and Services

The Geology Building (Plate 3.2), last in the row along University Avenue, was well designed by Bill Batt (the University Architect) and W. Pearce (Clerk of Works), implemented by ANU administrative staff (T. M. Owen and P. W. Brett), and built under the watchful eyes of geology staff members. Initially, deeper footings had to be sunk due to the ground structure and Sullivans Creek gravels. A central lecture theatre was surrounded by a 50-seat first-year laboratory, four smaller corner labs and four demonstrators' rooms. The entrance foyer had wall-map displays and two large display cases for museum and teaching specimens (Plate 3.3), plus an artistic polished-slab wall made from local rocks (Plate 7.4). Upstairs, a large, airy map-room over the lecture theatre was surrounded by staff offices, administration rooms and student cubicles with essential rock stacks. Also on this floor were a small Geochemistry Lab, X-ray analysis and diffraction labs, and a library. Technical services, the Lapidary Lab and a large museum with modern compactus units were located in an eastern single-storey wing. For a while, we also housed the Science Faculty Office, when

Geology at ANU (1959–2009)

Doug McAlpine was Faculty Secretary. Following public service custom, the staffrooms were shown on the initial plans to be different sizes depending on rank, but this was quickly changed when Professor Brown pointed out that all staff would expect to be promoted.

Plate 3.2 The Old Geology Building on University Avenue (now Engineering)
Photo: R. Barwick

The move into the new building was accomplished rapidly, although a spectacular accident occurred when bottles of ammonia and hydrochloric acid fell off a trolley next to the air intake; white fumes spread rapidly throughout the building, causing its rapid evacuation and a visit from the fire brigade (Plate 3.4). There was a problem with clocks installed in the corridors, which hung from the ceiling; several tall staff and students bent the stalks, so the clocks had to be raised. Another problem was with possums. One took refuge in the typewriter cupboard and another hibernated for a week in the X-ray machine. The small tearoom was a major social centre; a group of 10 contributed to purchase a tapestry print for $100 from Leo Seewen, the departmental photographer, who had a side business marketing artworks. This adorned the tearoom wall for many years. Several staff thought, rudely, that it was a most expensive tea towel. It disappeared during the move to the Botany Building!

The Buildings

Plate 3.3 Museum showcases in the foyer of the Old Building

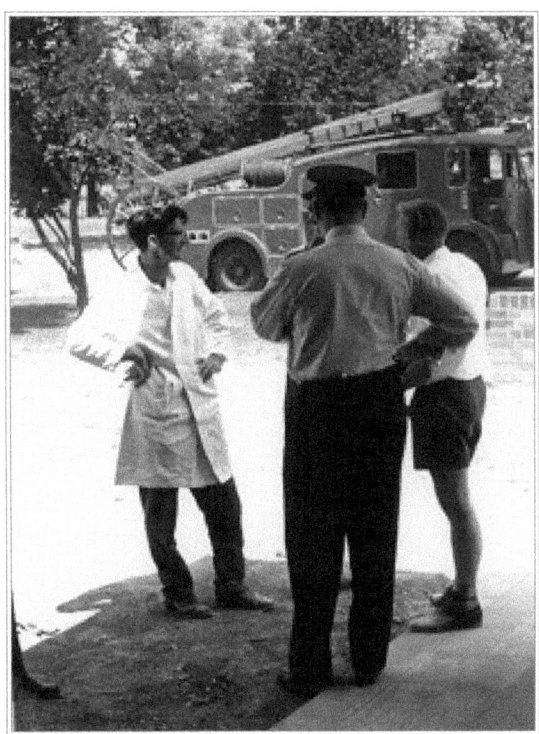

Plate 3.4 Firemen visit the Old Building. Alan White and George Halford discuss the situation with one of the firemen.

The eastern end of the grounds was swampy, so the Buildings and Grounds Department (now Facility Services) planted a small copse of oak trees to soak up the water. Later, in 1965, when planned extensions were to be built (Plate 3.5), there were protests against removing some of the trees. Moreover, the cost of the extension had been miscalculated so that 1.5 m had to be trimmed off the length of the building. The extension housed a 40-seat second-year laboratory, a well-lit library with small compactus for reprint storage, a palynology lab, a sedimentary lab and a large flume room, for demonstrating current action on sediments, in addition to three staffrooms and cubicles for graduate students. The roof of the new wing was none too watertight and some overseas students were amazed to see academic staff up to their ankles in water, mopping up after a storm.

Plate 3.5 The new wing of the Old Geology Building under construction
Photo: Sue Jephcott

The flume (Plate 3.6) required a large underground watertank outside the building. The flume was a joint venture with CSIRO; it was duly opened in 1970, by Sir Frederick White, Chairman of the CSIRO, who ceremoniously 'pulled the chain for the first flush'. The flume was an excellent teaching aid, but little used for research. There were two memorable projects: a physics honours student, Peter Killen, carried out a study on the design of surfboards, which was later published in the *Journal of Fluid Dynamics*; and an engineer from the

Military College, Duntroon, made a major study on the design of floodways in the Australian Capital Territory. Several years later, in 1982, the flume was dismantled and reassembled at CSIRO, after which the room was converted into a large geochemistry laboratory. This teaching and research accommodation served the department well for many years and made it a student-friendly base.

The 1991 forced transfer to the Botany Building (the New Geology Building; Plates 3.1 and 3.7) was a nightmare, but a major transport achievement. The X-ray machines, which were extremely heavy, and the large globe necessitated removal of doorframes. The valuable William Smith map in its large frame also required a skilled effort to move. The museum collection, which by then comprised several thousand specimens, had to be moved carefully and was split into two to fit the new storage space. We managed to take the compactus units with us and fitted them underground. We also acquired an outdoor storage cage for larger specimens. Unfortunately, the rock wall could not be moved, and is still anomalously decorating the Engineering Building (Plate 7.4). We managed to convince Buildings and Grounds that the old coconut mats in the upstairs entrance hall were dangerously worn, so we had them replaced with four rock slabs, chosen by David Ellis and provided by Hans Hensell. These are now used for teaching as well as for decorative purposes.

Plate 3.6 The flume in the Old Building
Photo: from K. Crook

The new accommodation was built around a central courtyard (Plate 7.4), with a large first-year laboratory on the ground floor and a large lab upstairs that was split in two with a sliding partition to give separate second and third-year labs. This arrangement enabled us to keep practical material for each class near at hand so that students could work in their own time.

Plate 3.7 The New Geology Building: the D. A. Brown Building
Photo: R. Barwick

In the New Building, staff and students shared a tearoom off the courtyard (Plate 3.8), and the students had a Geological Society office that housed a pool table. This room was later converted into an experimental petrology laboratory. Graduate-student and honours-student cubicles built upstairs were equipped with some of the furniture salvaged from the Old Building. When technical services and a large geochemical lab were built downstairs, the powerlines had to be drastically rearranged. For the first few years, members of the Botany Department continued to use the glasshouse on the northern side of the building; eventually (in 1994), it was removed and additional office space was added to house the growing Regolith Centre.

The Buildings

The unfortunate placement of a radioactive safe (lead-brick lined) under a cable channel in the thick concrete ceiling of the Geochemistry Lab caused a minor γ-ray radiation leak. Despite its trivial nature, the event occasioned considerable press coverage. The possible serious nature of this incident prompted a university-wide survey of the handling of radioactive materials.

There was no room for the map collection so this went on permanent loan to the Geography Department and the Australian Geological Survey Organisation (AGSO; now Geoscience Australia). The space allocated for it became a computer room, and a small drawing office was fitted out for marine survey work. The first seminar room was eventually converted to honours cubicles in 2001, and the small lecture theatre downstairs was rechristened the 'D. A. Brown Lecture Theatre' in honour of the first Professor of Geology.

Plate 3.8 The tearoom in the New Building on the occasion of Professor Campbell's seventieth birthday party. From left to right: Dick Barwick, Tim Munson, Robin Westcott, Ken Campbell, John Vickers, Judith Caton, Judy Papps, Wolf Meyer, Mike Rickard (standing), Prame Chopra, Judith Shelley, Dave Ellis, Des Strusz, Tony Phimphisane.
Photo: Gavin Young

The Buildings and Grounds Department had promised to repaint the building, but they never got round to it, so, then Head Technical Officer, Henry Zapasnik, organised a student working party to cover up the dirty battleship-grey walls with a bright coat of cream paint. The roof of the building had also sprung several leaks. This contributed to major flooding during a hailstorm in February 2007. The building has always been fully occupied and the influx of Postdoctoral Fellows required allocation of additional office space in the new Life Sciences Building next door. Plans to move the Centre for Resource and Environmental Studies (CRES) and to establish a closer link between Geography and Geology to make a School of Resource and Environmental Management (SREM) precinct have never materialised.

4
Teaching

Course Structure and Content

Initially, year-long courses (Geology I, Geology II and Geology IIIA) were offered, which included short sessions on different topics. More specialist units (Geol IIIB) were added in 1964. There was great flexibility of course content. Professor Brown took much of the responsibility for first-year teaching. Lectures and practicals were augmented by weekly films and television instruction items, and by numerous weekend field excursions. Some mapping practicals were conducted on the grass outside the building using wooden models of dipping rocks for students to measure and plot. Most staff also ran voluntary tutorials for small groups to assist with understanding course and practical work. They also helped integrate students into the department. These were continued with varying degrees of success for many years.

Cooperation with RSES was mostly good; Professor Jaeger insisted that all RSES non-geology graduates in geochemistry and geophysics took the Geology first-year course, and this kept our staff on their toes! RSES staff taught most of the geophysics courses for many years. Professor Anton Hales accepted our students' invitation to be patron of the students' Geological Society and attended their seminars regularly. Many RSES staff have contributed to lectures, and allowed staff and students access to their equipment and assisted with supervision of honours and graduate students.

Arthur Holmes' text on the *Principles of Physical Geology* was the basic first-year text in the early years. This was replaced with the magnificent *Perspectives of the Earth*—a text with an Australian emphasis produced for secondary school students by a team organised by the Academy of Science. It missed its intended mark; nevertheless, it was an excellent first-year text. With the recognition and acceptance of plate tectonics, many large American universities produced their own texts and the market was flooded.

Individual specialists shared teaching of second and third-year courses. Plate tectonics was a newly developing concept in the 1960s, replacing the continental drift hypothesis that was not widely accepted. Many geologists were 'fixists' and believed in island-chain connections between continents to account for similarity of fossils. Even our own petrologists were unconvinced, arguing that Professor Ted Ringwood's work on the mantle precluded drift! We first introduced the topic as lively evening seminars. Professor Brown was one of the few advocates in Australia (see Chapter 5). After several sessions, the students asked to hear the case against the theory, so we arranged seminars with outside lecturers. Professor Brown threatened to introduce one famous academician, a Fellow of the Royal Society (FRS), as 'fixed, rigid, and stationary'. Later, a new course on tectonics, taught by Mike Rickard, was added to the syllabus and an amateur film of plate movements was made in the first practical sessions. Students sabotaged the moving sea-floor-spreading model we had constructed by inserting a Playboy centrefold that emerged from the spreading ridge. Palaeo-magnetic and age-dating studies carried out in the Research School of Physical Sciences (RSPhysS) by Ted Irving, Ian McDougall, Ron Green, Don Tarling, Mike McElhinny, and others placed the ANU at the forefront of this new concept. From 1970 onwards, plate tectonics dominated tectonic theories and became the new, all-embracing paradigm. Together with an exploration boom, it gave a great boost to student interest and enrolments.

The introduction of the semester system in 1969–70 was useful for us, although it circumscribed our course structure. We experimented with different combinations. The 20-point degree comprised eight points in first year (normally four two-point units) and six points in each of second and third year (Table 4.1). This unit system allowed students failing in one unit to avoid repeating a whole year. Our problem was fitting all the introductory material into four second-year points; structural geology was paired at different times with sedimentology or economic geology until it finally settled in third year. Unlike some departments, Geology wished to keep two points in second year open for students to take other subjects in science (for example, chemistry) or even arts options. One student took Old Norse and ended up working for the Geological Survey of Finland. Another took Fine Arts.

Table 4.1 The 1990 course structure serves as an example of the Geology Degree course structure and the subject offerings for several years

First year	
Geol A11 The Earth Works (various)	Geol A02 Evolution Planet Earth (various)
	Geog A13 Landscape Evolution
Second year	
B03 Mineralogy (Eggleton)	B02 Palaeontology/Stratigraphy (Campbell)
B06 Sedimentology (Crook)	B04 Petrology (Ellis)
Option	Option
Third year	
C02 Structure/Tectonics (Rickard)	C01 Field Mapping (Rickard et al.)
C05 Geochemistry# (Chappell)	
C06 Geophysics# (Lilley)	C03 Palaeontology* (Campbell)
C04 Petrology* (Chappell)	
C07 Economic Geology (Walshe)	Seminar options C08 or C10 or C11, etc. (see below)

\# subjects offered in alternate years
* alternative options
Note: Staff leaders are in parentheses.

Student exam performance has generally been very good (Distinctions, 10–30 per cent; Credits, 10–50 per cent; Passes, 25–50 per cent; and Failures, 5–10 per cent). The introduction of a Pass 2 (P2) grade allowed poor students to continue their studies in other areas, rather than fail. Occasionally, some first-year performances were unsatisfactory, with up to 25 per cent failures. Our first-year course was subjected to a major review by the ANU Office of Research in Academic Methods (ORAM) in 1986 and it was well received. Faculty requirement for departmental committees (1970–80) was unpopular. Later, our students were invited to open staff meetings for a while; however, Geology maintained good contacts with students through labs and fieldwork.

In 1977, a general second-year course, Geology B09, was introduced to serve and attract Prehistory, Geography and other Arts or Economics students. It was dropped with the introduction of the general Geology A02 unit in 1980. Some joint courses with Chemistry and Zoology were also mooted, but only briefly accomplished—for example, the introduction of a second-year course in solid-state chemistry in 1983, which lasted for only five years.

Geomorphology had been accredited as a Science B point since 1980, and it was taught in Geography by two of our graduates: John Chappell followed by Brad Pillans. Several of our students took this unit. Initially in the Arts

Geology at ANU (1959–2009)

Faculty, Geography moved to Science in 1990, and, with Forestry, was later joined in a School of Resource and Environmental Science (SREM) with which we somewhat reluctantly associated. From 1990, we adopted a Y-structure with Geography for first year—with a first-semester joint course AO1 (The Earth Works) and separate second-semester offerings, AO2 (Evolution of Planet Earth) and A13 (Landscape Evolution). Several staff contributed to these lectures, but research priorities made it difficult to organise topics in logical order. Some students, especially from Geography, enrolled in Geology units with no high-school chemistry. To overcome this problem, Bruce Chappell gave a remedial crash course in introductory chemistry. Changes in first year resulted in dropping some of the more traditional material, such as crystallography and geological map interpretation, which had to be accommodated in later-year units. In 1996, students requested that lecture material be posted on the Web, so a Research Assistant, Colleen Bryant, undertook to transfer all first-year lectures.

Faculty criticism over offering courses to less than 10 students was countered by staff arguments that they needed to introduce their special expertise to students before honours year. In spite of heavy teaching loads, this system allowed most staff a semester free for research, and, by alternating some units, freed staff for occasional study leave. Specialist seminars were offered from time to time on topics including basin analysis (Tipper), marine geoscience (Crook), structural analysis (Rickard), vertebrate and theoretical palaeontology (Campbell and Tipper), surficial geology (Crook), Australian stratigraphy (Brown, Crook and Rickard), Australian sedimentary basins (Tipper), mathematical geology (Tipper), environmental mineralogy (Eggleton), thermodynamics and theoretical phase equilibria (Walshe), engineering geology (Rickard et al.), and petroleum and coal geology (Tipper). Field workshops to Heron Island and New Caledonia were run by Patrick De Deckker and later by Brad Opdyke. John Tipper was responsible for introducing computer work into the department with his basin analysis course and this was followed by a remote-sensing course taught by Prame Chopra. Half-point seminar units were largely abandoned by 1983. After 2004, when the department's name was changed to Earth and Marine Sciences, the course structure changed markedly (see Chapter 9).

Outside assistance was obtained for teaching some of the optional units. Dr Ian Williams assisted with the geochemistry course. The lack of a lecturer in the important field of geophysics has always been a problem and the course was run in third year, alternating with geochemistry, and both were assisted by outside lecturers. Members of RSES and AGSO (G. Bassi, J. Braun, I. Jackson, G. Davis, G. Houseman and B. Barlow) assisted with the geophysics course,

Teaching

and Ted Lilley (RSES) kindly took responsibility for many years. Ted Lilley's contribution deserves special mention as he ran or assisted with the course from 1977 to 1994, until Prame Chopra was appointed to the department. Prame was assisted by CRC-LEME Research Fellow Éva Papp in 1999–2006. Demonstrators for the course were graduate students from RSES (Dennis Woods, Steven Constable, Ian Ferguson, Nathan Bindoff, Richard Kellett, Graham Heinson, Paul Johnson, Dan Zwartz, Stuart Monroe and Michael Wingate). The research students who acted as demonstrators were a most important part of the course. They often brought in valuable ideas from the places from which they had come. Many also took away from the course ideas that they used subsequently when teaching geophysics elsewhere. Demonstrating was a valuable educational experience for them, and directly valuable on their own CVs. Of the students who took the course, many found geophysics enjoyable and satisfying, and later some were employed as geophysicists with oil companies, mining companies and government organisations such as Geoscience Australia (GA). A few became PhD graduates in Geophysics.

At the end of 1994, Ted Lilley's 'farewell gift' was to set up a Geophysics Prize for the students who came first in the course. For this purpose, the ACT Branch of the Australian Society of Exploration Geophysicists (ASEG) and the university established a prize to be awarded biennially (1996, then alternate years). In the event, when the course was held in successive years, the ASEG was pleased to fund the prize every year that the course was held (Geology student prizes and recipients are listed in Table 8.2).

Canberra proved a very good place to teach practically based geology courses. The large departmental tectonic globe (Plate 2) proved a great piece of equipment for the purpose of locating earthquakes from the records of the world's seismic observatories. A temporary pendulum set up in the department's stairwell (in both the 1970s and 1990s buildings) was used not only to measure gravity, but also to demonstrate, as a Foucault pendulum, the Earth's rotation. Outside the ANU, high-quality geophysical installations run locally by the ANU and BMR/AGSO/GA meant a wealth of places to visit to see geophysics in action. For example—visits were made to: the Canberra Magnetic Observatory, the Mount Stromlo Seismic Observatory and Heat-Flow Bore Hole, and the Black Mountain Palaeo-Magnetic Laboratory. The park-like grounds of the ANU and of Canberra in general provided excellent places for field measurements with geophysical instruments (generally provided by RSES and AGSO).

Other people contributed to expand our course offerings over the years. Max Gage, a visitor from New Zealand, gave a short course in Quaternary geology as early as 1963, but this did not become a popular part of the course until the 1980s. Roy Brewer (CSIRO) gave early soils lectures, and Sue Feary taught an introduction to statistics for the Australian geology course. Elizabeth Truswell (BMR), Donald Walker and Judy Owen (Research School of Pacific Studies/RSPacS) gave a course in palynology in 1980. During the late 1960s, Mike Plane (BMR) ran a course on Australian fossil mammals. Since 1978, Gavin Young (BMR, now ANU) has lectured and run practicals on early fishes in Ken Campbell's and Patrick De Deckker's palaeontology courses. David Ride lectured on fossil mammals. The advanced palaeontology course was offered in alternate years until the early 1990s (Table 4.1). Student field excursions in various years included trips to Forbes, Canowindra, Wee Jasper and Braidwood to study fossil fish, and the Monaro to study mammals.

Professor Shohei Banno, a visitor from Japan, assisted in teaching metamorphic geology. In 1971, the year Allan White left, metamorphic geology was replaced for a time by isotope geology taught by RSES staff. In 1974, the famous Germaine Joplin (RSES) taught the metamorphic geology course. Later, Wally Dallwitz (a retired BMR petrologist) helped with practicals in petrology before his untimely death in 1992. We had a hard fight to convince the Dean of Science that specialist courses could not be taught by other staff, and eventually we were granted a short-term Associate Lecturer (Warrington Cameron) to replace Allan White. Similar problems occurred later, especially with supervision of graduate students, as other staff left.

David Lock (a Postdoctoral Fellow) taught a course in lithofacies in 1987 while Keith Crook was on study leave. David Lock ran a course on industrial minerals. Similarly, Chris Jenkins stood in for Professor Campbell in 1981 while he was Dean of Science, and he made seven videotapes for introductory palaeontology. An engineering geology course was organised by Mike Rickard and taught with the help of several outside lecturers from industry and AGSO, especially Professor David Stapleton (Coffey/SA Institute of Technology), whose guided tours of the Snowy Mountains Hydro-Electric Scheme were most instructive and popular. With the appointment of Eric Best to the University of Canberra, this option was dropped and students were encouraged to enrol in engineering geology there. After Eric Best's retirement, Patrick De Deckker developed our first course in environmental geology.

In 1989, the ANU and the University of Canberra considered merging, and extensive discussions took place before the idea was dropped. In the meantime, the two geology departments had started joint teaching. Mike Rickard and Wolf Mayer taught a combined structural geology course successfully for several years. Teaching this course was made difficult by the differences in the structure and lengths of the courses at the two universities, which meant that the ANU students had to undertake extra work. Keith Crook and Graham Taylor taught sedimentology jointly, and John Walshe and Ken McQueen combined their economic geology courses. A highlight of the latter was a five-day 'magmatic-hydrothermal short course' that was run in the central west of New South Wales. The students were based in a caravan park in Cowra and a pub in Temora. Morning lectures were followed by visits to the Goonumbla (North Parkes) porphyry deposits, Browns Creek and Sheahan-Grants copper–gold skarns in the Blayney area and the Temora (Gidginbung) acid-sulfate-gold deposit. The final day examined the magmatic-hydrothermal breccias in the open pits of the Ardlethan tin deposits west of Temora. This short course gained much from the work of graduate students, and costs were defrayed by industry participation. University of Sydney staff and students also participated one year.

Despite having 13 postgraduate students in 1969, difficulty in attracting students in the 1970s endangered our staff positions. By 1979, we had only four honours students. In 1980, we had 39 first-year students, 18 second-year, 17 third-year and seven honours students. Note that by 2001 the number of honours students had increased to 16 (Plate 9.2).

The fourth-year honours course comprised a thesis based mostly on field mapping (50 per cent of marks), a seminar (10 per cent) and course work (40 per cent), of which Australian and world stratigraphy was a compulsory unit. David Brown, Ken Campbell and Keith Crook published in 1968 the first definitive text on Australian stratigraphy—*The Geological Evolution of Australia and New Zealand* (Pergamon, Oxford)—and, for many years, Professor Brown and Keith Crook taught this subject at third-year level. After they retired, the course was moved from third year to honours year and was taught by Mike Rickard and then David Ellis with several outside lecturers from AGSO and RSES, including one year by the visiting famous American tectonicist Warren Hamilton. This course was eventually dropped in 2008. Previously, a joint honours course with the Geography Department, as part of our SREM association, had given problems because the course formulae were different, as geography students were required to do less course work.

In the early days, a foreign language was required for honours and MSc candidates. Professor Brown ran a course in scientific Russian. There was also a compulsory exam on criticism of a published paper. After Professor Brown retired, these unpopular requirements were dropped. The honours year had a very heavy workload, especially as the date for thesis submission approached. The introduction of personal computers changed social habits somewhat as students no longer sought typing assistance. Honours graduates and their thesis titles are listed in Appendix 2.

In 1965, members of the department (David Brown, Ken Williams and Eric Conybeare) organised a national symposium on Undergraduate Geological Training to address concerns about the lack of practical work in topics related to future employment, especially in mineral exploration. Shortly after, a 1967 *Survey of Geoscientists in Australia* was published (K. A. Townley et al. 1968, Special Publication 1 of the Geological Society of Australia). Interestingly, this showed that many companies were happy to accept graduates with three-year degrees, as they provided on-the-job practical training for new employees.

In response to these concerns about practical training, together with the mining boom of the 1960s, the ANU Geology Department tried to introduce a professional four-year pass degree (like the Forestry Department). Faculty rejected this scheme, however, so for several years pass graduates not eligible for honours were encouraged to undertake a Graduate Diploma. A merit pass for the diploma allowed entry to an MSc course. This also acted as a convenient means of assessment of potential overseas MSc students. Three students from the Geological Survey of Indonesia took advantage of this scheme. Sae-un Hardjoprawiro worked in Fiji for his MSc; Sufni Hakim proceeded to Royal Holloway College, London, for an MSc; and Priharjo Sanyoto worked with the BMR team in Kalimantan.

We had some altercations with the Australasian Institute of Mining and Metallurgy (AusIMM) in the 1990s. They criticised our lack of applied subjects and wished to accredit the ANU courses. The department refused to do this, as we maintained that our coverage of basic science prepared our students well for a wide variety of earth science careers, and many mining companies were keen to employ ANU graduates (see comments by alumni, Chapter 11). Moreover, in 1981, the Geological Society of Australia (GSA) had set up a professional body, The Australian Institute of Geoscientists, that could accredit geology graduates for professional work. By now there were two geology departments in Australia (at JCUNQ and the University of Tasmania) that offered MSc courses in exploration geology for students who wished to gain more applied training. Subsequently, the Department of Earth and Marine Science (DEMS) has been accredited by AusIMM (see Chapter 9).

The staff also contributed to outside teaching. Professor Brown, Dr Conybeare and Dr Rickard, in turn, ran an introductory course for the ANU adult education program, until 1981. George Halford, the Museum Curator, ran an adult education course in gemology for several years. Professor Campbell arranged short courses for school science teachers when the Wyndham Scheme added geology to the NSW high-school curriculum in 1962. Ken Campbell and Mike Rickard also contributed an article on the state of geological teaching to *The Australian Bicentennial Project Bulletin* in 1987–88. The department hosted introductory sessions for prospective high-school students and CSIRO summer schools. Dick Barwick (then teaching in Zoology) ran a laboratory-based school program introducing fossils to primary school students as part of the 1988 Science Festival. Bruce Chappell and Keith Norrish (CSIRO) ran an X-ray course for industry for several years. Patrick De Deckker commenced the 'University of the Sea' program with colleagues from Sydney University, which has so far run two cruises involving 20 honours and postgraduate students on the French research vessel *Marion Dufresne*. Patrick was also involved with Peter Hancock (Visiting Fellow) and Sarah O'Callaghan (HTO) in running a student 'Mineral Ventures Program' for high-school students, which featured visits to local mining sites in New South Wales, as well as visits to Geoscience Australia (GA) and the University of Wollongong. In September 2006, Patrick organised a 10-day visit for 20 students and staff from Tokyo University. They made field excursions to Yass and the South Coast, where they were based at the ANU's Kioloa field station. Bear McPhail has presented short courses on hydro-geochemistry and environmental geology to the Universities of Tasmania, Monash and Princeton, and the Minerals Tertiary Education Council. Mike Rickard gave short courses in tectonics at James Cook University in 1970, and the Geological Survey of Indonesia at Bandung, in 1989.

As could be expected with the complex matters of balancing teaching and research loads and study leave, staff meetings were occasionally acrimonious. Especially difficult were reluctance and differences of opinion over course changes, the forced move to the new building and budget allocations. Mike Rickard, as Head of Department, reported to the 1990 Review Committee that he viewed his job as akin to the 'governor' of a steam engine—attempting to stop the balls flying off in all directions!

Geology at ANU (1959–2009)

Student Fieldwork

Plate 4.1 Manduramah field excursion, 1968. Back row: Keith Crook, Alex Grady, Barry Webby (Sydney), Professor David Brown, Nick Arndt, Pat Davoran, Peter Madsen. Middle row: Ken Campbell, Leonie Chalker, Gavin Young, Vitas Labutis, Penny Simpson, Ian Powell. Front row: Wal Bucknell, Mike Huleat, Neil Williams, Anne Felton, David Bigg.
Photo: David Moore

Early field excursions (1962) were combined with those of the University of New England (courtesy of Professor Alan Voisey) and the University of New South Wales, visiting localities in the Orange–Wellington area (Plate 4.1), New England, Yass and Tilba Tilba. Most courses included weekend fieldwork. Initially, Professor Brown insisted on taking first-year students to camp— mostly at Taemas in mid-winter. He also made a habit—based on his navy training—of waking everyone up at 6 am with a cup of tea. This worked well as a hydrostatic alarm clock! On one camp, a student caught a black snake that he intended taking back to CSIRO Wildlife, but it escaped from the sack and caused pandemonium in the bus. Fortunately, no-one was hurt and the snake escaped. Mark O'Connor (the poet, then an arts tutor living at Burton Hall, ANU) asked to come on a first-year field excursion to Lake George. He amused us all by demonstrating his judo skills, throwing female students neatly over barbed-wire fences—a trick that led to ripped jackets when others tried it!

Plate 4.2 First-year excursion, 2003. David Ellis and 'Bear' McPhail instruct students at the ANU's field station at Kioloa, New South Wales.

Plate 4.3 David Ellis poses in front of the famous unconformity at Myrtle Beach
Photo: Gavin Young

Geology at ANU (1959–2009)

When student numbers exceeded 50, we switched to a joint camp with Geography in the Snowy Mountains, led by Geoff Hope. Later, Henry Zapasnik rented a ski lodge in the off-season for an introductory Kosciuszko excursion. The 8 km round-trip walk to Blue Lake was a good test of stamina for budding geologists. There was also a weekend coastal excursion for first-year students run by Tony Eggleton and other staff members.

Plate 4.4 Students mapping at Bermagui
Photo: Stephen Cox

Second-year palaeontology classes regularly visited Ulladulla with Ken Campbell to study the major unconformity between Ordovician greywackes and Permian glaciogenic sandstones and Permian fossils (Plate 4.3). The structural geology course examined the famous complex structures at Bermagui with Mike Rickard and more recently with Stephen Cox (Plate 4.4). Broken Hill was a common destination for petrology, economic geology and structural geology courses (Plates 4.5 and 4.6). One of the first trips led by Ken Williams started with a spectacular breakdown in sight of the town. The oil in the VW Kombi van boiled dry, seizing the engine. It turned out that paper towel from the last oil check had been left in the engine compartment and this had blocked the air intake! This was some introduction for Shohei Banno, a Japanese visitor; but later the poker machines completed his education of Australian bush life.

Plate 4.5 1968: Neil Williams leading his group underground at Broken Hill. From left to right: David Bigg, Dave Christie, Dick Price.

On another trip, led by John McDonald, the students had been invited to a dinner in the Town Hall by AusIMM and the mine managers. Since they were camping, no-one had smart clothes, but by accident Mike Rickard had packed a dozen ties, so the students managed to appear reasonably presentable. Petrology students still tour the fantastic geology of Victoria and Broken Hill with David Ellis in third year (Plate 4.5). On one occasion, Ellis arranged for the Mayor of Broken Hill, Peter Black, to address the students at their final-night dinner. The Mayor delivered a risqué speech that insulted the female students, but Megan James (a geology/law student) took him to task in her reply. To make amends the following day, Black took some of the students out to see the International Sculpture Park and the famous double-fold locality. David Ellis has even organised camel rides to cap off the week-long stay in Silverton.

Geology at ANU (1959–2009)

Plate 4.6 Relaxing after fieldwork: students at the Silverton Pub, Broken Hill, ca 1994. From top left: 1. Matthew Adams, 2. Elisha Ahern, 3. Alan Cunliffe, 4. Greg Miles, 5. Chris Allen, 6. Leah Moore, 7. Allison Britt, 8. Louise Mitchell, 9. Clinton Rivers, 10. Claudia Camarotto, 11. Doug Eramus, 12. Paul Ferguson, 13. Stuart Girvan, 14. Chris De-Vitry, 15. Tony Phimphisane, 16. Warwick Crowe, 17. Stuart Love, 18. Kath Barr, 19. Ennis the barmaid.

The earliest departmental major field-mapping camps were run alternately as hard-rock and soft-rock 10-day camps in the end-of-year vacation. For example, one studied granites (hard rock) at Moruya and Dalgety (1966) before settling on Berridale as the study area; the other worked on the fossiliferous limestone (soft rock) at Taemas and Boambolo. During a camp at Taemas, a Visiting Fellow, George Grindley, was sent out with a mapping party. At the end of the day, the students returned exhausted, for nobody had told them that George was a NZ alpine explorer. In 1963, two nuns took geology in preparation for their science teaching under the new NSW Wyndham Scheme. Climbing fences in their habits was a problem, so Professor Brown had a portable stile made. This proved to be too clumsy, so the students made a human stile for the nuns to leap over the fences. At night, as they had to be transported to the nearest convent, they were driven back to Yass by Ken Campbell in the vehicle, renewing our beer supplies.

With the advent of the semester system in 1969–70, all fieldwork had to be related to a course and assigned points. The petrology camp was shortened to five days and was combined with the mineralogy course. Initially, John McDonald ran this camp at Cobargo. After he left the department, Tony Eggleton with George Halford, Ian Smith and other staff ran these camps at Tilba Tilba to map the Mount Dromedary volcanic centre (Plate 4.7). Ultimately, a scientific paper was produced as well as a pamphlet and map of the Golden Volcano for local tourists. Numbered plaques were erected around the mountain to guide tourists to the geological features.

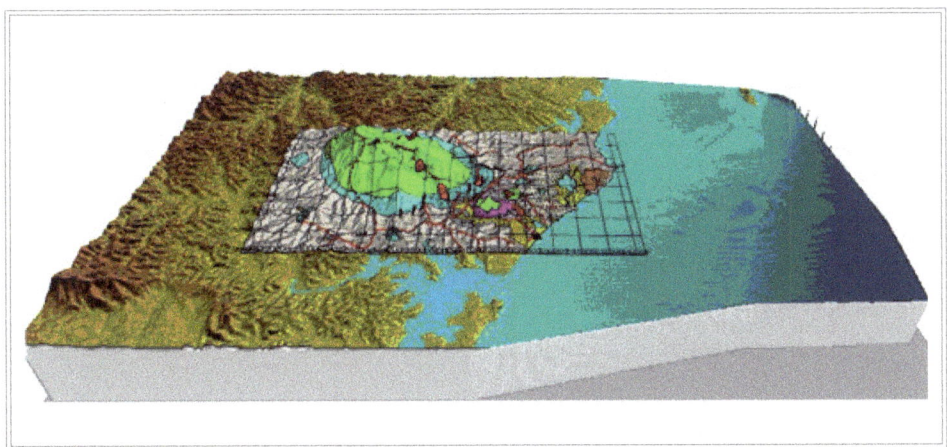

Plate 4.7 The Golden Volcano, Tilba Tilba
Photo: Patrick De Deckker

The third-year stratigraphy camps remained at 10 days with one point credit (Geology C01). These were held in several different localities throughout New South Wales (Plate 4.8), with different staff leaders—including: Boambolo (Crook, Feary, Opdyke); Nundle (Crook); Goodhope (Campbell, De Deckker, Rickard); Quidong (Campbell, Rickard, Opdyke); Michelago (Rickard); Tathra–Bunga (Rickard, Tipper); Rangari (Tipper); Eden (Crook, Rickard, Opdyke); Carcoar (Rickard, Walshe); Boorowa (Rickard, Opdyke); Wee Jasper (Opdyke, Rickard, Young); Gowan Green (Opdyke, Strusz); Wellington–Burrandong (Opdyke, Strusz); Broken Hill (Lister). Property owners and managers have always been helpful and amenable to having students working over their land and we acknowledge in particular Ken Kilpatrick, the property manager of Cavan Station, and Ian and Helen Cathles at Wee Jasper, who have allowed us access over many years.

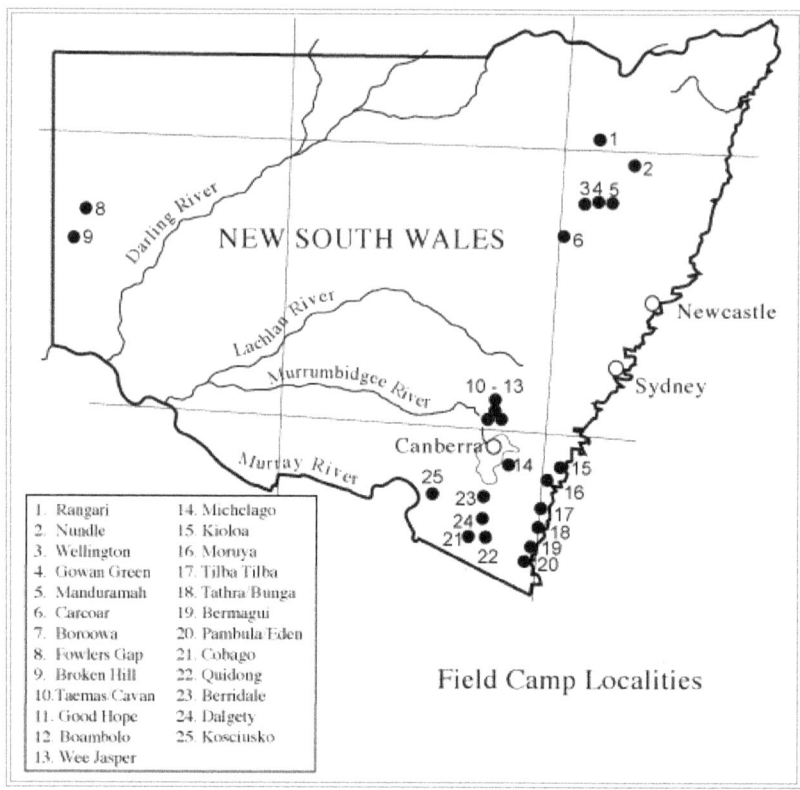

Plate 4.8 Map of field camp locations in New South Wales

Apart from good mapping training, these camps provided many memorable events. While mapping at Moruya, Tony Taylor showed students how to trace granites through unexposed fields by noting hornblende in the anthills, and Allan White found a new use for a G-pick when wading across part of Tuross Lake—we had a good feed of fish that night (Plate 4.9).

At Dalgety in 1968, we mapped sequential granite plutons. For evening entertainment, we organised a dance, but the band did not turn up, so it was a flop. The bar was kept open late by one of the demonstrators (Bob Day), while students flirted with the publican's daughter. Then we were woken in the early hours of one morning by some irate local parents looking for their daughters. We denied all knowledge of their whereabouts!

During the camps at Rangari, we mapped Permian volcanics, and were treated to roast suckling pig courtesy of Henry Zapasnik, who had been out hunting the night before with the property owner. At Carcoar, we camped in a B&B run by two charming ex-actors, and two students were attacked and stomped on by a wild emu while mapping Ordovician volcanics and a granite pluton.

Teaching

Plate 4.9 Alan White 'fishing' in Tuross Lake
Photo: Sue Jephcott

Students mapped the Quidongan unconformity and Silurian volcanics at Michelago. Here we camped in shearers' quarters. Each night we visited the Bredbo pub for showers and to play on the small hexagonal pool table. Greg Harper, the draughtsman, turned up to barbecue a whole sheep for our last night's campfire sing-along. This became a tradition, with Henry Zapasnik repeating the effort at Good Hope camps. Here, we used small boats to cross the dam to map Devonian limestones. Tim Munson even managed to get marooned as one of the boats slipped its moorings and miraculously floated back to the caravan park.

At Mountain Creek (Taemas), where we mapped Devonian sediments and volcanics, Radi Popovic took to wrestling a large goat, which so enjoyed the exercise it made it difficult for students to run from the cookhouse to the sleeping quarters. During a camp at Good Hope, we were visited by Herman Jaeger and David Skevington, two world-famous graptolite experts. Herman went on to compare and correlate the eastern Australian faunas with those at type localities in Germany. The last Eden camp to map Devonian volcanics and sandstones was marred by coastal development, as we had to obtain permission from more than 200 property owners to enter their land! This camp, held in mid-winter, was based in caravans with map plotting in the open air. Naturally, the nearby Seahorse Inn with warming whiskey macs became a substitute base each evening. For several years, we mapped the spectacular Devonian limestones at Wee Jasper, on the property of Ian and Helen Cathles (Plate 4.10). We took advantage of low water levels in Burrinjuck Dam to trace stratigraphic units.

For three years (1995–97), the CO1 field-mapping camp held at Boroowa, New South Wales, was run jointly with the University of Canberra (led by Ken Mc Queen) and with the ANU Geography soils course (led by John Field). This cooperation between the bedrock and surficial student groups was most beneficial. Students also presented their results to the townsfolk at a popular evening seminar in the local pub. Late one night, some students caused a disturbance by climbing into the swimming pool and running about naked. The local cop, who was in his first week of duty, was not amused and remonstrated angrily with John Field, the camp leader. For the last-night dinner, students bought dress-up clothes ($10 maximum) from the local 'op-shop'. The elderly female shop assistant was somewhat perplexed by the cross-dressing, but she was even more surprised when all the clothes were donated back the next morning.

The CO1 (field mapping) course included a short laboratory-based course on photo interpretation, for which Mike Rickard had taken, in 1973, a photo-grammetry course with the Australian Mineral Foundation in Adelaide. Also, for many years, a field day of plane-table mapping gave students a basic idea of surveying. Although old fashioned, this exercise was always enjoyed by students. One student (Garry Davidson, now lecturing in Tasmania) was so enthusiastic that he swam the sighting staff out to small offshore rocks to complete the map at Picnic Point.

Teaching

Plate 4.10 Field camp at Wee Jasper, 2005. Back row, left to right: Taylor Walsh, Jane Thorne, Helen Tait, Jenna Roberts, Mitchell Bouma, Graham Nash, Andrew Tankey, Tim Curran, Brad Opdyke, Brian Spall, Malcolm Mann, John McDonald, Claire Bennett, Max Collett, Jennifer Burke, Joshua Knight. Front row, left to right: Melissa Jinsberg, Rhiannon Mann, Alice Menyhart, Nicholas Rankin, Peter Meadows, Meryl Larkin, Antonia Bigault.
Photo: from B. Opdyke

For several years, voluntary excursions were also run to Heron Island as an optional seminar course. These trips were led by Patrick De Deckker with assistance from Tim Munson and Henry Zapasnik. The students contributed to ongoing research on coral reefs. These trips were continued later as a seminar offering under Brad Opdyke's leadership.

An extra camp was run in 1967 to accommodate several students who had missed the main camp. A small group, with Mike Rickard and Brian Chatterton (a PhD student), mapped at Fowlers Gap in western New South Wales (with permission from the University of New South Wales, which owned the property). The planned early start was delayed by the non-appearance of one student, who was rushed into the Geology carpark in full evening dress still asleep. She was bundled into our van, waking up several hours later. The map the students produced on this trip was used for many years as an exercise for the photo-interpretation class. On returning through Broken Hill, two students (the Arndt brothers) made a bargain purchase of some Pro Hart paintings.

John Tipper has written about fieldwork:

> ANU was great experience—to see geology from a Gondwanan perspective and from an old and stable craton, to see how teaching and research can be put together in a (usually) positive way, and to have confirmed the fact that even the most magnificent of modelling efforts can crumble in front of some apparently insignificant little outcrop. Experience is what the geologist needs above all, preferably in the field with students. Sometimes, however, experience can be exasperating, as Ken Campbell found out during a Taemas camp. I have changed the student's name but his interaction with Ken Campbell was as follows.
>
> Theo belonged to a religious group who claim the Earth is 5000 years old—plus-or-minus 10 minutes…Ken, of course, was determined to show this young man the error of his ways, and to do so in an impeccably scientific way, based on field observations we could all make together. A soil profile was found and Ken asked how long it had taken to form: *'This must have taken at least 3000 years, Theo, what do you say?' 'Yeah', Theo agreed. Five minutes later, just round a track bend, were two such profiles occurring together, the second clearly superimposed on the first. '3000 years for the first profile, Theo, then 3000 years for the second?' 'Yeah!' 'So how old is the Earth, Theo?'*—from an understandably triumphant Ken. '5000 years', said Theo!

Most honours students made a major study of a field area for their theses and generally these were done with their own transport. Many of these projects were carried out in areas of interest to staff research, however, some honours students carried out research projects in the laboratory supervised by RSES staff. Staff have also been involved in supervising graduate research–student fieldwork in many locations throughout Australia—in several cases, attached to BMR/AGSO field camps. In addition to the scheduled fieldwork, students have organised their own end-of-year excursions with staff assistance. These are reported in Chapter 8.

There have been only one or two vehicle accidents on field trips over the 50 years; in one, a car was rolled and a trailer was jackknifed. The only serious accident occurred in 1973 on the return from a Tasmanian excursion when a vehicle collided with our van and Graham Taylor was hospitalised for more than a year with a compound leg fracture. A minor collision also occurred in the Hunter Valley. Allan White is reputed to have avoided a collision by swerving and driving full-speed through a service station! Field accidents include a cliff fall, scalded legs and a dislocated shoulder. There have been several snake encounters but fortunately none serious. On one Taemas trip, John Brush, an avid cave explorer, jumped into a sinkhole on top of a snake; he was pulled up to safety by his mates, with his boots covered in venom. Several students have had painful contacts with electric fences—none serious, however.

5
Research

Staff Research

1. Palaeontology

Professor David Brown

Professor David Brown was a world authority on polyzoa. His work done at Imperial College and the British Museum of Natural History was published as a treatise on New Zealand Tertiary Cheilostomatous polyzoa (Bryozoa). He also published (Royal Society Victoria) a major study on the Tertiary polyzoa of Victoria. He was invited to study samples from Antarctica and drill cores from Bikini Atoll made before atomic tests. He also carried out pioneering palaeolatitude studies with Ted Irving (RSES), plotting Labrinthodonts and other reptiles in their palaeolatitude positions rather than their present latitudes; and showing that the palaeo-positions gave better fits to the expected equatorial zoning. This major early contribution to the tectonic-plate theory was read as a Presidential Address to the ANZAAS convention held in New Zealand and published in the *American Journal of Science* in 1964. Professor Brown also made several translations of Russian geology texts, especially books on the Russian kimberlites and diamond pipes by Nick Sobolev, who became a Visiting Fellow at the ANU. He also translated Russian palaeomagnetic results for Dr Mike McElhinny (RSES) and an oceanographic text for which Professor Stewart Turner (RSES) acted as technical editor.

Professor Ken Campbell

Professor Ken Campbell first worked on Permian brachiopods from the Bowen and adjacent basins. This work was published in international journals and set a framework for several other authors in Australia and New Zealand. It also provided a basis for contributions to the *Geology of Queensland*, published by the Geological Society of Australia (GSA) in 1960. While at the University of New England, Armidale, his study of carboniferous faunas from the Hunter

Valley to Bingara was published in an attempt to understand the stratigraphic sequence in New South Wales. A Nuffield Dominion Fellowship allowed him to take a sabbatical year at the Sedgwick Museum, Cambridge, UK, and resulted in BMR's *Bulletin 68* on 'Terebratulid brachiopods' of the whole Australian continent.

The trilobites at Yass first attracted Professor Campbell's attention, and, in 1965, he spent a year with Professor Whittington at Harvard on a Fulbright Fellowship studying trilobites. This resulted in two major works: a bulletin published by the Oklahoma Geological Survey, and a joint bulletin of the Museum of Comparative Zoology at Harvard. Papers on the trilobite and brachiopod faunas of the Canberra, Yass and Taemas regions were also published. This also opened up research on the eyes, muscle scars and other features of phacopid trilobites that was published as the Clarke Lecture of the Royal Society of New South Wales in 1975.

A chance discovery of a second specimen of the dipnoan (lungfish) of the genus Dipnorhynchus at Taemas began a completely new line of research (I am told on good authority that Ken jumped and whooped for joy on making this discovery: MJR). In 1970, Ken led an expedition to the Kimberley in north-western Western Australia that included David Brown and George Halford together with Dr Alex Ritchie from the Australian Museum. They collected limestone nodules at Gogo that contained remarkably well-preserved fossil Devonian fish. This led to an extensive research program, which continues to this day.

The need for a meeting of workers on early vertebrate evolution resulted in Professor Campbell organising a meeting in Canberra and Sydney. About 40 people from around the world assembled and, after papers were presented in both venues (Plate 5.1), members were shown Australian fossil beds at Forbes, Taemas, Wee Jasper and Cobar. The results of this symposium were published in 1984 in a single volume by the Linnean Society of New South Wales.

Research

Plate 5.1 International Symposium of Vertebrate Palaeontology at the ANU, 1983. Back row, left to right: Alex Ritchie (Australian Museum), Moya Smith (Guy's Hospital, London), Peter Forey (Natural History Museum, London), Charles Marshall (ANU, now at Harvard), John Long (now at Victoria Museum), Bob Jones (Australian Museum), Emilia Vorobjeva (Academy of Sciences, Moscow), Anne Kemp (University of Queensland), Jim Warren (Monash University), Mahala Andrews (Royal Scottish Museum, Edinburgh), Ken Campbell (ANU), Daniel Goujet (Natural History Museum, Paris). Front row, left to right: Richard Lund (Adelphi University, New York), Pan Jiang (Geology Museum, Beijing), Elga Mark-Kurik (Institute of Geology, Tallinn, Estonia), Hans-Peter Schultze (University of Kansas, Lawrence), Chang Mee-mann (Institute of Vertebrate Palaeontology, Beijing), Richard Barwick (Zoology, ANU), Wang Nianzhong (Institute of Vertebrate Palaeontology, Beijing).
Photo: Gavin Young

Alone and in collaboration with Dick Barwick (Plate 5.2), Ken Campbell has published more than 40 papers on the evolution, palaeoecology and phylogenetics of fossil lungfish. This body of work is of international significance, and resulted in a period in London working on the histology of teeth, and in Chicago at the Field Museum on a Visiting Fellowship. Several overseas fossil-fish experts including Moya Smith and Mahala Andrews from the United Kingdom have visited and studied at the ANU with this group of palaeontologists. This work has also opened up an investigation of new designs in biological morphology by a genetic process of gene regulation, and Professor Campbell has collaborated with Professor George Miklos in this research.

Plate 5.2 Ken Campbell and Dick Barwick on a fossil-collecting trip at Wee Jasper

2. Micropalaeontology

Professor Patrick De Deckker

Professor Patrick De Deckker is a micropalaeontologist who thrives on multidisciplinary research. He has studied many aspects of the continental and marine realms, including micropalaeontology, sedimentology, the geochemistry of biogenetic carbonates, palaeolimnology and climate change. His research has extended from lakes to oceans, and he has led eight marine research cruises at sea. In collaboration with students, postdoctoral fellows and colleagues, Patrick has studied the following marine groups: acantharians, calcareous nanoplankton, diatoms, dinoflagellates, ostracods, planktic and benthic foraminifera, and radiolarians. The Micropalaeontology Laboratory at the ANU is a world-leading centre. Patrick pioneered the study of trace-element chemistry of ostracod shells with Allan Chivas and Mike Shelley (RSES). This technique is now used internationally, especially for the reconstruction of past climates.

The combination of trace-element and stable-isotope analysis on ostracods was further explored with Alan Chivas, Mike Shelley, Steve Eggins and Aleksey Sadekov (RSES), and with Ulysses Ninneham at Bergen University.

A new project, supported by ARC funding, aims to 'fingerprint' (geochemically and microbiologically) aeolian dust in the Australian region. There is international interest in this research, with participants from the Max Planck Institute of Marine Microbiology, Bremen University, RSES and the Medical School at the ANU, and Monash University. Patrick's work has yielded more than 160 scientific papers, as well as several journal volumes and books. Additional research undertaken by Professor De Deckker is reported in the section on the Australian Marine Quaternary Program in Chapter 6.

3. Petrology

Professors Allan White and Bruce Chappell

Professors Allan White and Bruce Chappell (Plate 5.3) organised and supervised a major research program for many years during which they mapped and sampled most of the granites of the Lachlan Fold Belt (Plate 5.4). This study also included contributions from a large number of undergraduate students. Dr Keith Norrish of the CSIRO Division of Soils had developed the world's most advanced methods of X-ray fluorescence analysis (XRF) and Bruce Chappell was given the opportunity to use these methods in his laboratory at the ANU. Over the years, this laboratory developed a strong international reputation. Bruce was heavily involved in the early stages of development of automation of XRF equipment, and a mechanical sample changer was awarded a Certificate of Merit in the Prince Phillip Prize for Australian Design in 1971 (Plate 5.5). The geochemical labs were set up twice in the Old Building and then in a larger room in the New Building. The laboratory analysed not only a large number of granite samples over the years, but also a wide variety of other rocks—most notably, samples from all of the Apollo lunar missions. Bruce Chappell was honoured with the Mawson Medal and Lecture by the Australian Academy of Science in 2003, and in Japan by a special issue of *Resource Geology*, 'The Chappell Volume' (2006).

The work of White and Chappell yielded some major concepts of granite genesis. Their restite model proposed that the variations in composition within many granite suites resulted from variations in the amount of residual solid material from the source rocks entrained in the melt. Such granites are now referred to as 'low temperature' and contrast with the 'high-temperature' granites that form from complete melts, such as the Boggy Plain Pluton near

Geology at ANU (1959–2009)

Adaminaby, which was studied by Doone Wyborn. The presence or absence of old zircon is critical in understanding these two types of granite. This work was done in collaboration with Ian Williams (RSES), who had earlier made an honours study of the granites north of Jindabyne. White and Chappell showed that the two distinct groups of granites that were first recognised in the Berridale region of New South Wales owe their differences to derivation from the partial melting of igneous and sedimentary source rocks. These I-type and S-type granites are now widely recognised. Bill Collins, a former ANU student, has challenged many of these conclusions and a lively debate has ensued. For his honours thesis, Bill Collins studied A-type granites—the third type recognised both in south-eastern Australia and more generally.

Plate 5.3 Professor Chappell (centre) after the award of his DSc in 1990, standing with his original supervisor and research collaborator, Emeritus Professor Allan White, and former graduate students Richard Price, now Dean of Science and Engineering at the University of Waikato, New Zealand (left), and Ian Smith, University of Auckland (right)

Photo; D. Ellis, from ANU Geology Department Annual Report, 1990.

Research

Rick Hine, another of Chappell's students, recently published a major paper with Bruce on the Cornish granites. Bruce supervised students working on Queensland and Malaysian granites. Much of the work on granites was supported by mining companies through the Australian Mineral Industry Research Association (AMIRA) (see Chapter 6). Bruce Chappell also assisted with work for the Department of Prehistory, ANU, and for Dr G. Ward of the Institute of Aboriginal Studies, analysing cherts and materials deleterious to Aboriginal paintings.

Plate 5.4 Bruce Chappell modelling the latest in makeshift wet-weather gear to study granite
Photo: D. Wyborn

Geology at ANU (1959–2009)

Plate 5.5 XRF analyser with modern automatic sample changer
Photo: M. Rickard

Dr Warrington Cameron

Dr Warrington Cameron continued his early work on ophiolites in Greece and Cyprus while at the ANU, as well as studying oceanic volcanics and boninites in New Caledonia. He was a member of an international collaborative project on komatiites. In 1980, he supervised PhD students Donal Windrum, working on metamorphic rocks in the Northern Territory, and Dave Walker, in Papua New Guinea.

Professor Richard Arculus

The prime focus of Richard Arculus's research has been the study of island arc-backarc systems, such as the Lesser Antilles, Kamchatka–Kurile, Izu–Bonin–Mariana, New Britain–Manus Basin, Solomon Islands, New Hebrides–North Fiji Basin, and Tonga–Lau Basin, with emphasis on primary magmas, fractionation histories, and overall mass fluxes at convergent plate margins. This work is part of a broader program (in collaboration with many other researchers) directed to understanding the origin and evolution of the continental crust, the processes leading to the establishment of major terrestrial structural and geochemical domains (core, mantle and crust), and fluxes between these domains. In part, this research led to an interest in the possible contamination of the mantle by surficial oxidised materials and the nature of the redox balance of the mantle. As an advocate of basalt as the volumetrically most important parental magma in arcs, Richard has pursued answers to the consequent problem of why the continental crust has a bulk andesite composition. A possible locale for the geochemical complement of an upper felsic (that is, granodioritic) continental crust is a mafic lower crust. On the basis of studies of 'xenolithic' lower continental crustal suites, the absence of expected distinguishing characteristics of an upper crustal complement have become clear: delamination and recycling in subduction zone systems for these complementary materials are required. In the past few years, microanalysis (by LA-ICP-MS) used with the explosive products of intra-oceanic arc systems has developed the first comprehensive record of Izu–Bonin–Mariana arc evolution. This has been achieved through studies of ash layers recovered by deep-sea drilling, and initiating studies of melt-inclusions in the Taupo eruptive sequences. During the past five years, Professor Arculus has participated in and led several marine research voyages targeting submarine arc-backarc systems and recovering glassy materials essential for the pursuit of studies of dissolved gases and potentially volatile trace elements.

Studies of the petrologic, geochemical and tectonic evolution of island-arc systems have been pursued at the ANU in collaboration with: staff members Professor David Ellis, Dr John Mavrogenes and Dr Joerg Hermann; Postdoctoral Fellows Steve Eggins and Ian Parkinson; Visiting Fellows C. Ballhaus, D. Gust (Queensland Institute of Technology), A. Kersting (Lawrence Livermore National Laboratory), R. Duncan (Oregon State University); and graduate students E. Chen, L.-Q. Cao, C. Bryant, T. Teng, C. Spandler, H. Patia, C. Qopoto, T. Berly and J. Brownlow.

Professor David Ellis

Professor David Ellis's research has concentrated on metamorphic processes involved in the evolution of the deep Precambrian crust of Australia, Antarctica, Sri Lanka and China. In Antarctica, David studied with Japanese colleagues from the National Polar Institute, Tokyo, Professor Zhao and Dr Shiraishi, and Professor Hiroi (Chiba University), and with graduate students David Young, Jonathon Kilpatrick and Warwick Crowe. They studied areas around the Mawson and the Lutzar–Holm Bay regions and Prince Charles Mountains, concentrating on the granulites and charnockites. They concluded that the latter were a new distinctive igneous magma type rather than metamorphic, as normally assigned. Such rocks were derived by the partial melting of dry granulitic crust at high temperatures. This group also recognised previously unknown volcanic derivatives. Another result was the recognition of a 500-my-old pan-African belt in eastern Antarctica that extends into Sri Lanka. Together with Dr Shiraishi and Dr Sheraton (AGSO), David also studied the geochemistry of the Yamoto syenite.

This work was extended to Sri Lanka with Professor J. Berg (USA) in a study of carbon dioxide streaming in lower crustal xenoliths, and with Dr Chiraisi, Dr Hiroi and Dr Mark Fanning (RSES) correlating the geology of the Lutzow–Holm area in Antarctica with that of Sri Lanka, extending to southern China in a study of late Proterozoic ophiolites with Dr Malcolm McCulloch and Chinese workers. He has also worked on ultra-high-pressure rocks from the Tianshan region of western China with Professor L. Zhang (Visiting Fellow from Beijing University); and Samantha Williams has carried out a honours project on subducted carbonates, recording evidence of extreme depth of burial.

Work in the Musgrave Ranges of central Australia with PhD student Makenya Maboko (RSES) recognised retrograde eclogites cutting granulites and unusual felsic volcanics, indicating that these rocks originated deep (35–40 km) in the crust. David also worked on partial melting of the crust at Cooma, New South Wales, with Dr Obata, a Visiting Fellow from Kunamato University, Japan. In connection with this research, David has participated in several international conferences. In 1998 in Leningrad, David Ellis was a keynote speaker on ultra-metamorphism and anatexis. At the Symposium on Enderby Land at the National Polar Institute, Tokyo, he was joint convener of Symposium 31 on 'Deep crustal structures of orogenic belts and continent: seismic and geological models', and also at the twenty-ninth International Geological Congress in Kyoto, Japan. Together with Bruce Chappell, Professor Ellis helped to organise the International Association of Volcanism and Chemistry of the Earth's Interior (IAVCEI) conference held in Canberra in 1993. As Project Leader of the International Geological Correlation Project (IGCP) 236 (Precambrian events in Gondwanaland), David organised field workshops in Kenya and Sri Lanka. He has taken study leave at the National Polar Institute, Tokyo, and at the University of Maryland, where, with Professor Mike Brown, he visited the Adirondack Mountains to examine charnockites and granulites.

In addition to field studies, Professor Ellis has set up a high-pressure/temperature laboratory in the New Geology Building at the ANU to work on crustal melting. He has carried out many experiments with many collaborators, including

- partial melting of mafic rocks of the deep crust and subduction zones with Professor A. B. Thompson (ETH, Zurich)
- experiments on Pb-Th-U diffusion in zircons from eastern Australia, with Dr J. Lee and Ian Williams (RSES) and Gladys Warren, a Visiting Fellow
- high-pressure experiments on ilmenite-rutile equilibria and trace-element partitioning
- high-pressure study of Fe-Ni-Co-S melts in equilibrium with olivine, with C. Ballhaus of the Max Plank Institute
- sulfide equilibria in high-grade metamorphic terrains, with R. Frost, University of Wyoming, and John Mavrogenes (RSES)
- trace-element compositions of garnet—important for Nd-Sm dating of garnets—with Dr Steve Eggins (RSES) and Professor Hiroi
- experimental studies of the AlF content of titanate and zircon-mineral equilibria with Postdoctoral Fellow Ulli Troitzsch and Dr Andrew Christy. This work led to a patent on zirconium titanate material for semiconductor use.

He has also made petrological descriptions of 'oven stones' from Papua New Guinea for Professor Jack Golson (RSPacS). At present, David is working with Sara Beavis on the study of the chemical effect of recycled water and sewage effluent on coastal systems.

4. Sedimentology and Marine Geology

Keith Crook

Keith Crook, after completing his honours in 1953, commenced a study of the Sydney Basin lower Triassic strata, which he submitted for an MSc. Early in 1956, he started fieldwork south-west of Tamworth, which was the basis for his PhD (awarded in 1958 by the University of New England). Keith then spent a few months at the University of Melbourne before taking up a National Research Council of Canada Postdoctoral Fellowship at the University of Alberta, from September 1959 to April 1961. He visited many sites in Alberta and British Columbia, and made one trip to the eastern United States (Connecticut, New Jersey and Pennsylvania). His visit to Alberta resulted in some graduate students as well as staff members and visiting fellows coming to the ANU. Leaving Canada in April 1961, en route to the ANU, he visited Colorado, Arizona, southern California, Hawai'i (Big Island and Oahu), Fiji and New Zealand.

Keith Crook's research interests at the ANU were initially focused on sedimentary petrology (later expanding to sedimentology and tectonics). Initially, he studied the evolution of the Tasman Geosyncline (as it was called in the 1950s and 1960s). In 1956, he began petrological studies of sedimentary rock samples colleagues had collected from South-West Pacific countries. These led, in 1961, to a paper on diagenesis in the Waghi Valley sequence in Papua New Guinea, and, in 1963, to a paper on burial metamorphism in Fiji. These were precursors to on-going major involvements in Pacific Island countries and their territorial waters.

Keith also had two minor, but recurrent research strands that commenced early in his career. One, beginning in 1956, termed 'surficial geology' or 'landscape evolution', was based on his exposure to soil science at Sydney University. The other, commencing in 1957, was his interest in aspects of structural geology as preserved in weakly or early deformed rocks. At this time, too, he collaborated with Eric Conybeare in producing a useful *Manual of Sedimentary Structures*, published by the BMR. Two other recurrent strands emerged later. His identification of shatter cones at Gosse Bluff (Northern Territory) led to an interest in astrogeology and high-energy catastrophic events that has recently been renewed. Completion in the 1960s of a BA majoring in Political Science led to his participation in the development of the Australian Labor Party's science policy, and, in the 1970s, to his involvement in the Society for Social Responsibility in Science (ACT).

Early in May 1965, John Chappell arrived from New Zealand to take up a PhD scholarship at the ANU to begin studies in Papua New Guinea. By 8 May, John and his co-supervisors, Keith Crook and Eric Bird from the ANU Geography Department, were examining sites around Lae and the Markham Valley. Finding little dateable material, they flew north-east over the Rai coast of the Huon Peninsula to inspect the remarkable terraced topography in kunai grassland, which they had noted in wartime oblique aerial photos while in Port Moresby. These turned out to be a series of uplifted coral reefs, which formed the basis for John's PhD thesis (awarded in 1973). This research revived the Milankovitch hypothesis of episodic Quaternary glaciations and sea-level fluctuations. This field trip marked the beginning of Keith's continuing fieldwork in Papua New Guinea and the South-West Pacific, which continued during his term as Science Program Director in the Hawai'i Undersea Research Laboratory, University of Hawai'i at Manoa, from June 1992 to June 2004.

By the late 1960s, the interrelationships between stratigraphy, sedimentology and tectonics were topics of growing interest nationally and internationally. Keith took the opportunity on his first ANU sabbatical leave in 1967 to spend three months in the Soviet Union. This was followed by visits to southern Wales, Alberta, Colorado, Texas (as a visiting lecturer), southern California, Kauai (Hawai'i), Tahiti and then Otago, Canterbury and the west coast of the South Island of New Zealand. Observations and contacts arising from these visits provided input into his publications.

His interest in New Zealand geology and its relationship to that of eastern Australia had arisen during the preparation in the mid-1960s of the book *Geological Evolution of Australia and New Zealand* with colleagues David Brown and Ken Campbell. This led to several papers (both theoretical and integrative) published during the following 15 years that examined the relationships between tectonics, mafic magmatism and sedimentation in ancient and modern island arc settings. Much of this work was based on his supervision of honours and graduate students mapping in the Tumut area of New South Wales.

In 1980, Keith's research interests 'grew webbed feet'. He became Secretary (and later Vice-Chairman, Chairman and Past-Chairman) of the Consortium for Ocean Geosciences of Australian Universities, which promoted Australian participation in the Ocean Drilling Program. Five research cruises on US (Hawaiian), Japanese and Russian research ships in Papua New Guinea and Solomon Islands waters resulted, providing data and insights for his research projects. These cruises also led to his participation in the South Pacific Applied Geoscience Commission (SOPAC) and its Science, Technology and Resources Network (STAR). After his move to Hawai'i in mid-1992, Keith completed five more cruises and served as chair of STAR until October 1999.

In September 2002, an unsuccessful application for an ARC Discovery Project on the Cainozoic tectonic evolution of northern Papua New Guinea led to Keith's renewed interactions with former ANU colleagues and, ultimately, in July 2004, to his appointment as an ANU Visiting Fellow, based in the by then renamed DEMS. His research since then has been conducted largely in conjunction with his wife, Dr. E. Anne Felton (see Chapter 11), on the sedimentology of rocky shorelines. While in Hawai'i, they became interested in the origins of cobble and boulder-gravel deposits on modern and ancient shorelines of the Hawaiian

Islands. This led to a series of joint papers in *Sedimentary Geology* on the sedimentology of rocky shorelines. This interest in an until recently neglected aspect of sedimentology has led to an ongoing joint research program on the boulder deposits of the South Coast of New South Wales, at Little Beecroft Head (Jervis Bay), outer north head Bitangabee, and the boulder-beach ridges at Leatherjacket Bay. Keith's work has resulted in 120 articles, two books and 126 conference abstracts.

Brad Opdyke

Brad Opdyke's research has concentrated on oceanic and carbonate deposits in northern and north-western offshore Australia, together with Kriton Glenn (honours student), studying the chemistry of shelf-water and surficial sediments on the Sahul Shelf (Ashmore Reef). Papers on this work, and on the shrinking coral–algal habitats during the last glacial maximum and its impact on the carbon cycle, were presented in Sydney to the Australian Marine Science Association. Brad has participated in several marine cruises, including

- a Quaternary Marine Science cruise under Dr De Deckker's leadership, carrying out water analyses from Darwin to Fremantle
- the 1996 AIMS cruise off the west coast of Australia to obtain a detailed picture of the sediments north of Exmouth Gulf and west of Barrow Island
- blue-water palaeoceanographic research off Scott Plateau—work presented to the fifth International Palaeoceanography Conference in Halifax, Canada.

He is an invited member of active international teams investigating 'Land–ocean interactions in the coastal zone' and 'Future directions of biochemical cycling' research. In 2000, he was awarded an ARC grant for a 'Palaeoceanographic study of the western Pacific Warm Pool and its interactions'.

5. Basin Analysis

John Tipper

While at the ANU, John Tipper's research concentrated on quantitative and theoretical stratigraphy, with emphasis on computer-based modelling; also of significance (once the university had been persuaded to part with what was then an inordinately large amount of money to purchase its first high-end graphics workstation) was his design and development of some sophisticated dynamic visualisation software—in effect, the first geological flight simulator. Collaborative research took place mainly with Tony Eggleton and John Walshe (on diagenetic patterns and processes in the Denison Trough, Queensland). Rais Ahmad joined this project as a Postdoctoral Fellow. John developed an industry collaboration project with Wilkinson Seismic (Chris Wilkinson is a former student of the department). David Kelley, a Visiting Fellow, was employed on this project to develop seismic interpretation software.

Doctoral students supervised by John Tipper were Michael Swift (with whom he spent a memorable three weeks on the *RV Franklin*, initially measuring submarine heat flow on the Exmouth Plateau, then finally chasing the tail of a cyclone), Patrice de Caritat (who studied the Denison Trough), Chris Pigram (jointly supervised by David Falvey from the BMR), and Chunping Ding ('the master of the workstation'). Doctoral supervision was started for Ejazul Haq, Trent Liang and Nasre Sobhan, but this was relinquished when John left the ANU. John was joint supervisor for Peter Hill during research for his MSc. Honours students he supervised included Jim Arthur, Saxon Palmer, Cam Schubert and Melissa Fellows, and Graduate Diploma students supervised (some jointly) were Peter Sharman, Phil Jankowski and Arnel Mendoza.

6. Structural Geology

Stephen Cox

Stephen Cox's research at the ANU since 1998 has focused on exploring aspects of the coupling between deformation processes and fluid flow at depth in the Earth's crust. This work has involved high-pressure, high-temperature experimental studies on rock-analogue materials and field-based studies of fluid–rock interaction and deformation processes in faults, shear zones and related vein systems. The work has involved extensive microstructural and microchemical studies, and has been complemented by theoretical and numerical modelling studies. The field-based research has been conducted in both non-mineralised and mineralised settings. The focus of the field-based research has been to develop an understanding of fracture-controlled flow regimes and the impact of fluid processes on the mechanical behaviour of rocks at depth in the crust. The work has been applied to explore structural controls on localisation of mineralisation from deposit scale to regional scale. The experimental and microstructural studies have aimed to explore and quantify the effects of reactive pore fluids and dissolution-precipitation creep processes on the strength and mechanical behaviour of rocks, especially the mechanics of fault zones. The research has also quantified the roles of deformation and fluid-rock reactions in influencing the fluid transport properties of deforming rocks.

Much of Stephen's fieldwork at the ANU has dealt with Archaean mesothermal gold systems in the Kambalda region of the Yilgarn Craton, with funding provided by the ARC Linkage program, Western Mining Corporation (WMC) Limited, Gold Fields Limited and AMIRA International Limited. A number of projects at Kambalda have supported various honours students (J. Alexander, M. Ison, S. Zasiadczyk, R. O'Leary, C. Nicholson, A. Hickey), as well as PhD students K. Ruming (at the University of Newcastle) and M. Crawford. Stephen also collaborated with PhD student P. Nguyen at the University of Western Australia. These deposit-scale projects have refined understanding of how ductile and brittle deformation processes localise fluid flow and reaction in hydrothermal systems, and influence the location and geometry of ore shoots. In particular, the research has highlighted the dynamic nature of stress states, fluid-pressure states, and flow and reaction in episodically rupturing fault systems in the upper crustal seismogenic regime. As such, the research has provided insights into the mechanical behaviour of fault zones and the role of fluids in controlling fault mechanics near the base of the continental seismogenic regime. A key additional result is that ore formation in such regimes is far from a 'steady-state' process. Rather, fault regimes are dominated by large and cyclic permeability change, consequent episodic fluid flow, associated changes in flow velocity and pathways, and incremental formation of ore deposits via hundreds to thousands of flow events.

The research has shown how the geometries and locations of fault stepovers and bends influence fluid pathways and permeability anisotropy in fault-controlled flow systems. The research has also shown how episodic failure and associated changes in fluid pressure and hydraulic gradients in fault zones can influence the relative roles of fluid–rock interaction, fluid-pressure drops and fluid mixing in driving ore deposition at various stages in repeated fault-valve cycles.

A key feature of many Archaean gold deposits in the Kalgoorlie–Kambalda region—and indeed many other Archean gold systems globally—is that deposits tend to be localised in small displacement faults and shear zones up to several kilometres from the kinematically related, high-displacement, crustal-scale faults and shear zones. Furthermore, deposits tend to be clustered in groups, with each cluster (or goldfield) about 20–30 km apart along particular crustal-scale fault systems. Analysing the gold-hosting fault systems near Kambalda in a seismogenic framework led to the recognition that the ore deposits are localised in host faults that would be regarded as aftershock structures in a modern seismogenic context.

Subsequent work, in collaboration with postdoctoral student Steven Micklethwaite (2002–08), and funded by the ARC Linkage program and AMIRA International Limited, has developed the application of 'stress-transfer modelling'—used elsewhere in aftershock risk analysis—to understand controls on the distribution of goldfields in crustal-scale fault systems.

Experimental studies of mineral analogue materials, such as biphenyl, with Postdoctoral Fellow Jürgen Steit (1998–2000), and funded by an ARC Discovery Grant, allowed the analysis of microstructural processes in situ in a see-through microscope stage, during progressive deformation involving competing brittle and dissolution-precipitation creep processes in artificial fault zones.

Work on fluid–rock reaction processes in networks of shear zones in the Mont Blanc and Aar Massifs in Europe provided Stephen Cox with an opportunity to work on spectacular outcrops among equally spectacular scenery, but in very challenging topography. Stephen was ably guided through some very precarious locales by Postdoctoral Fellow Yann Rolland (2001–03), who had completed his PhD at Grenoble and, usefully, was a fully trained French alpine guide, as well as being an excellent structural geologist and geochemist.

Field-based and stable-isotope studies of vein systems in the Devonian carbonate sequence at Taemas (New South Wales), with honours student C. Wilcox, have demonstrated the operation of a major fracture-controlled hydrothermal-flow system during regional crustal shortening in the area. Vein formation was associated with upward flows of modified meteoric fluids. Reactive transport modelling of spatial and time variations in $\delta_{18}O$ compositions of veins indicate that, although the flow system was active throughout the protracted folding history, flow was episodic rather than continuous. Individual pulses of fluid flow were likely associated with episodic breaching of an over-pressured fluid reservoir by repeated fault-rupture events at depth.

ARC-funded experimental rock deformation studies at RSES, especially in collaboration with Postdoctoral Fellow Eric Tenthorey (RSES, 2001–07) and RSES PhD student Silvio Giger (2004–07), allowed analysis of time-dependent changes in fault strength, mechanical behaviour and permeability evolution.

Other projects with honours students include studies of quartz-vein arrays on the South Coast of New South Wales by A. Shepherd (jointly supervised with Mike Rickard); vein formation associated with the Inglewood Fault at Gympie, Queensland, by B. Witham; and controls on the development of gold mineralisation at Adelong, New South Wales, by C. Mitchell (jointly supervised with John Mavrogenes).

7. Structure and Tectonics

Mike Rickard

Mike Rickard (Plates 5.6 and 5.7) has carried out mapping studies of the structure of orogenic zones in Donegal Eire, the Canadian Appalachians, Fiji and Norwegian Caledonides, and with honours and graduate students in central Australia, the Canberra region, the South Coast of New South Wales, and in Tumut. In 1967–68, one of the first Australian Research Grants Committee (ARGC) grants funded an expedition to the Patagonian Andes, with Dr Kerry Burns of Macquarie University, to collect palaeomagnetic samples for testing the oroclinal theory to explain the bend in the southern Andes. During study leave in 1969, Mike sat in on the famous MSc structure course given at Imperial College London by Professors John Ramsey and Neville Price, and brought back valuable teaching specimens and photos of the currently fashionable 'double folds' in Scotland.

Plates 5.6 A geologist at work. Mike Rickard examining folded rocks with cleavage structures at Bermagui during the SGTSG International Conference.

Research

Plate 5.7 Mike Rickard 'examining' undeformed sediments at Merimbula

Mike worked on the *Tectonic Map of Australia* as the executive secretary, and attended an international workshop in Montevideo in 1967. He also convened the Australian WG9-Geodynamics project that produced the first structural section across the Lachlan Fold Belt from Wagga Wagga to Batemans Bay. Much of the fieldwork for this project was carried out by Research Assistants Bruce Duff and Peter Ward. In 1976, Mike participated in the Swedish Geodynamics Project by mapping in the western Caledonides of Norway. In that year, Mike also revised earlier fieldwork in the Canadian Appalachians for eventual publication by the Canadian Geological Survey. Some local consulting work has involved searching for and testing aggregates for quarry operations, and studying the structures of mines at Peelwood, New South Wales, and with Ken McQueen at Cowarra near Bredbo.

Publications include five regional map bulletins and other regional structural studies. He has also presented work on cleavages, dating of cleavages, quartz-vein arrays, monoclinal kink folds, pluton-spacing geometries, polygonal arrays, and orogenesis on an expanding Earth. Mike recently assisted with the publication of a GSA field guide to the geology of the Canberra region.

8. Economic Geology

Ken Williams

Ken Williams developed techniques for the use of electron-microprobe and atomic-absorption spectrophotometry for major, minor and trace-element geochemical studies of ores and related gangue minerals. The research focused on hydrothermally zoned mineral assemblages around the Heemskirk granite near Zeehan on the west coast of Tasmania (Plate 5.8). On one field trip in 1964, Ken complained because he could not rent a VW 'beetle', which he wanted as the Morris 1000 provided could not drive up the railway lines and this meant walking miles further! On the return journey to hand in the car, the lights were 'repaired' by a local garage mechanic throwing a bucket of water at the car to wash off the mud.

In 1967, Ken undertook an experimental study of the Ni-Sb-S system with G. Kullerud at the Geophysical Laboratory in Washington. Research on the geochemistry of a variety of ore deposits was continued after he left the ANU to take up a post at Stanford University.

Plate 5.8 Ken Williams on fieldwork in western Tasmania
Photo: from Ken Williams

John McDonald

John McDonald worked on various mineral deposits in New South Wales, including consulting work on the old Peelwood deposit. He supervised PhD students Ron Britten, Doug Mason and Joe Whalen working on mineralisation in Papua New Guinea and island arcs of the South-West Pacific.

John Walshe

John Walshe's research with numerous students on the mineral deposits of New South Wales concentrated on hydrothermal processes and wall-rock alteration. He developed a geothermometry system for chlorite with Research Assistant Brian Harrold (1982), whose early tasks were to develop computer programs to construct redox-pH phase diagrams and compute the physico-chemical conditions of chlorite growth. John still maintains a version of the former program at <http://ems.anu.edu.au/phase/phase.html>. Subsequently, Brian's role evolved to that of the departmental guru on all things computing and his skills have also been well used by other departments and schools.

John also carried out sulfur-isotope analyses together with a Postdoctoral Fellow, Stewart Eldridge. John continues his research on mineral deposit processes with the CSIRO Exploration and Mining Division in Perth. As Chief Research Scientist, he is working on mineral-system models of Australian mineral provinces that take account of the flux of volatiles/metals from the deep Earth and explicitly link resource-forming processes to other large-scale spatial and temporal processes through Earth's history. This research is done in collaboration with Geoscience Australia, State surveys and universities through cooperative research centres.

John Mavrogenes

John Mavrogenes was awarded more than $2 million in government and industry grants for his work on sulfur solubility in silicate melts, gold and copper solubility in high-pressure/temperature fluids, the effect of metamorphism on ore deposits, the origin of opals and the composition of subduction-zone fluids. Besides laboratory work, this has entailed fieldwork at Broken Hill, New South Wales, the Gawler Craton (the Challenger Deposit) in South Australia and at Lightning Ridge, New South Wales. Since 1996, John has supervised 10 graduate students and 27 honours students. One of John's students, Andrew Tomkins, who completed his PhD in 2004 and is now a lecturer at Monash University, has recently won the Lindgren Award from the Society of Economic Geologists for the best young economic geologist.

9. Mineralogy

Professor Tony Eggleton

Throughout his academic career, Professor Tony Eggleton has focused his research on the layer silicates. One branch of research, which he pursued from his PhD until the early 1990s, was the study of the so-called modulated layer silicates. His work on this mineral group has been collaborative, with Professor Steve Guggenheim (University of Chicago), who became a Visiting Fellow in the ANU Geology Department for two years. These silicates are minerals with complex structures and sub-micron crystal size. Determination of their structure required the development of new methods for silicate structural analysis. From this work, the atomic structures of stilpnomelane, ganophyllite, minnesotaite, parsettensite and greenalite were elucidated. Dr Angela Heinrich, a Postdoctoral Fellow, also contributed to this research when she solved the problem of the structure and polytypism of bementite.

Clay minerals formed the second branch of Tony's research, which was directed towards understanding how clays formed as primary minerals weathered. This research involved several postgraduate students, particularly Wang Qi-Ming, Jill Banfield, Mehrooz Aspandiar and Watcharaporn Keankeo. Postdoctoral Fellows Kath Smith, Maïté le Gleuher, David Tilley and Ian Robertson also contributed greatly to these studies, during which the group discovered the mechanisms of mineral weathering and clay formation for most of the rock-forming silicates—namely, feldspars, olivine, garnet, pyroxenes, amphiboles, biotite, muscovite and chlorite. These studies led to wider investigations into regolith geology, mainly through the research of honours and graduate students.

For the past 15 years, Tony has concentrated his research work on collaborative investigations with Graham Taylor, an ANU PhD graduate and University of Canberra professor, principally into the formation of the Weipa bauxite, in northern Queensland. In 1988, Eggleton and Taylor established a cross-campus research centre: the Centre for Australian Regolith Studies (CARS). This centre, funded by the ANU and ARC, employed Doctors Le Gleuher and Lock and enrolled several graduate and honours students. A Masters in Regolith Studies by course work and sub-thesis was established as well as an undergraduate regolith geology course. In 1996, CARS was one of three research groups (with the AGSO Regolith Group and CSIRO Division of Exploration and Mining, Perth) to win Cooperative Research Centre funding, allowing the establishment of the Cooperative Research Centre for Landscape Evolution and Mineral Exploration (CRC-LEME) (see Chapter 6 for more detail).

10. Environmental Geology

D. C. 'Bear' McPhail

'Bear' McPhail's most significant contributions to research have been in the fields of hydrothermal geochemistry and regolith geoscience. His main focus has been on measuring mineral solubilities, identifying important metal aqueous complexes and deriving the thermodynamic properties of reactions over wide ranges of pressure, temperature and composition. The metals studied include tellurium, copper, iron, zinc and gold. In particular, he has contributed to understanding mineral solubilities and aqueous complexes under highly saline conditions. Without these kinds of contributions, geochemists and other scientists are limited in understanding how elements are transported through the Earth's crust. Some of Bear's best studies so far have been on the hydrothermal geochemistry of tellurium, and subsequently on copper chloride and other complexes—the last leading a team at Monash University. Many of his ideas were catalysed by Dr Joël Brugger, now an ARC Professorial Fellow at Adelaide University.

Since moving to the ANU in 2002, Bear has focused mainly on regolith geoscience, as part of the CRC-LEME. His contributions include the effect of salinity on the sorption of metals on iron oxyhydroxides, the speciation and solubility of gold under regolith conditions, and the solubility and stability of zinc alteration minerals. Notably, another part of his research has been to study the effect of microbial activity on gold mobility, driven by recent PhD student Frank Reith and resulting in a 2006 paper in *Science*. At the ANU, Bear also built a research team, including Research Fellows Dirk Kirste and Sue Welch and several PhD and honours students, all focused on mineral exploration or environmental problems related to the regolith, including groundwater geochemistry in mineral exploration, salinity dynamics, and sulfur budgets in inland acid-sulfate soils. Much of this only recently completed work is now in preparation for publication. Bear has had 20 journal publications and 49 conference abstracts since 2001.

11. Oceanic Geochemistry

Michael Ellwood

Michael Ellwood's research focus has two major thrusts.

1. Understanding the cycling of trace metals in the upper ocean and their interaction with phytoplankton: changes in global climate resulting from increases in atmospheric carbon dioxide concentration coupled with anthropogenic activities represent a massive scientific and humanitarian problem. Michael's research aims to evaluate the synergistic links between trace metals (iron, zinc, cobalt, cadmium) and macro-nutrient cycling in Southern Ocean waters and to determine how they influence primary production and community structure. Understanding how these elements influence primary production and their links to community structure is paramount to developing appropriate responses to preserve the biodiversity and ecosystem structures that underpin the marine food web.

2. Developing and applying new proxies and approaches to reconstructing past oceanic environments, with a particular focus on reconstructing the historical distribution of silicon, which is required by diatoms and sponges to build their siliceous skeletons. In the present-day ocean, diatoms account for 40 per cent of primary production, which makes them a key taxonomic group in modulating the sequestration and export of carbon dioxide from the surface to the deep ocean. In the past, however, atmosphere carbon dioxide concentrations and climate have fluctuated significantly, leading to warm (interglacial) and cold (glacial) periods. Only through understanding the factors that could lead to changes in diatom production, and hence the ability of the ocean to draw down carbon dioxide, can predictions of future climate change be made.

Michael has been actively involved in teaching courses at second and third-year levels on the chemistry of the oceans and marine biochemistry for the Bachelor of Global and Ocean Sciences and BSc degrees, and he is supervising several honours and PhD students involved in these research efforts.

Table 5.1 Research assistants (in approximate date order)

D. Moss	S. Love	K. Barr
L. Belbin	G. B. Alexander	C. Rivers
E. McPherson	D. Kelly	P. King
J. Heaslip	D. Holloway	G. L. Deacon
D. Ryan	R. Affleck	L. Maconachie
R. Tenge	J. D. Woodhead	D. B. Tilley
J. Bein	M. M. Hedges	C. Newham
K. Smith	N. Fraser	R. Skirrow
W. Kiene	C. J. Johnston	C. Wilkinson
I. Martinez	J. R. Sellar	A. Watchman
C. R. Marshall	P. N. Porecki	S. Bygrave
B. Duff	C. Rumble	R. Musgrave
P. Ward	C. E. Pratt	C. Camarotto
E. J. Reid	C. Findlay	D. Franklin
M. Glikson	P. English	C. Mee Mann
J. Stanner	V. Urbaczewski	C. Ma
Q. M. Wang	D. Champion	W. E. Chen
A. Raisuddin	S. Veitch	J. Zhao
M. Le Gleuher	V. Elder	I. W. MacIntosh
P. Burkle	J. Shelley	A. Barrie
A. Shepherd	A. Davis	C. J. Bryant
B. Harrold	V. Drapala/Passlow	I. Roach
D. M. Long	D. N. Young	
S. M. Matysek	W. Crowe	
K. M. Tuffin	C. L. Moore	
M. Chorley	L. Dansie	

Research by Postdoctoral Fellows, Visiting Fellows and Research Assistants

Over 50 years, the department has hosted many short-term researchers. Research Assistants, Postdoctoral Fellows (Tables 5.1 and 5.2) and Adjunct Professors have been employed mostly on research grants, and some of their work is included with staff research. The first two Research Assistants were appointed to balance the technicians working in geochemistry and petrology. Lee Belbin provided computer assistance to Dr Rickard and Dr Crook, and Daphne Moss worked in palaeontology with Professor Campbell and Professor Brown. The number of Research Assistants and Research Fellows increased dramatically as the staff became successful in attracting research grants. In the 1970s, there were four grants; there were 10 in the 1980s, increasing to more than 20 from the 1990s onwards. Many honours, graduate and postgraduate students were also employed for periods of up to a year as Research Assistants on external research grants. It was difficult to gather accurate information on all of these positions and there could be omissions.

Postdoctoral Fellows (Table 5.2) were also employed on research grants or were supported by special fellowship awards, especially in the major projects AMIRA, GEMOC, CRC-LEME and Marine Science (see Chapter 6). The work of several of these Fellows is recorded later in this section. Until recently, Research Fellows were not required to formally teach, although several did voluntarily. With the loss of Demonstrator (Associate Lecturer) positions, the department resorted to using PhD and honours students as demonstrators for practical classes.

Many of these positions gave new doctoral graduates their first chance of tough, and sometimes insecure, employment.

Table 5.2 National Research Fellows and Postdoctoral Fellows

D. Lock	A. E. Rathburn	K. Evans
S. Lonker	J. X. Zhao	R. Ahmad
J. Long	F. Vander Hor	A. Heinrich
C. S. Eldridge	T. Esat	C. Heinrich
P. Blevin	S. Eggins	I. Parkinson
C. Allen	T. Ulrich	T. T. Barrows
P. Pridmore	I. G. Morrison	C. L. Moore
H. D. Hensel	M. Le Gleuher	G. Taylor
B. Musgrave	I. Martinez	F. Gingele
P. Wells	P. E. Mathe	É. Papp
M. Ayress	J. Magee	J. E. Streit
J. Sellar	G. Young	J. M. Caton
C. Perkins	W. Qiming	S. Welch
I. Borissova	X. Sun	D. Kirste
S. Nees	J. Reeves	M. Lenahan
I. D. Campbell	C. Martin	
I. Robinson	C. McFarlane	

Bob Musgrave

Bob Musgrave is a good example. He came to the ANU in March 1988 and, for more than a year, assisted with research on different projects, mixing his time with geophysical modelling of the Tumut Trough and the Woodlark triple-junction, and with trying to date Monaro laterites. Keith Crook was offered a ride from Canberra to Melbourne in Alan Bond's experimental blimp, to demonstrate its use as a platform for airborne gravity measurements. As Keith was unable to go, he persuaded Bob Musgrave to make the trip. Bob then held a succession of Postdoctoral Fellowships—first at the University of Tasmania (1989–91), and then with the Ocean Drilling Program at Texas A&M University (1991–93). He was then appointed as a lecturer at La Trobe University until the Geology Department there closed in 2003. He then survived in a fixed-term fractional position as Senior Research Fellow at Monash University, and also as a part-time Lecturer in Geophysics at La Trobe University, Melbourne University and RMIT. After a few months of unemployment, he was appointed a Senior Geophysicist with the NSW Geological Survey and a conjoint Fellow with the University of Newcastle. Another example of career difficulties is reported by John McGee in Chapter 11.

Visiting Fellows fall into different categories, ranging from retired geologists and casual visitors to overseas experts who spent several weeks or months working with staff on specific projects. In Appendix 1, Visiting Fellows are listed only once to avoid repetition; however, those who have had long-term appointments or made several visits to the department are marked with an asterisk. Again, it was difficult to give an accurate listing, as several visitors were casual or unofficial and do not appear in the department's annual reports; apologies to anyone omitted.

For several years, Professor Brown encouraged visitors from Russia, and he made several translations of their work, especially on ultra-mafic rocks and diamonds. These visits ceased after Russia invaded Afghanistan and the ANU cancelled the exchange arrangement. Groups of Chinese and Japanese geologists followed. A group of retirees from ANU/BMR/AGSO/GA made up probably the largest group of palaeontologists in the country, if not the world, and they have produced an abundance of publications. This group dubbed themselves the GEROC (Geriatric Old Codgers) (Plate 5.9). Several international and Australian visitors have worked with this group of fossil-fish specialists.

Plate 5.9 The GEROC (Geriatric Old Codgers) logo designed by Dick Barwick for the group of retired palaeontologists working in the Geology Department

Plate 5.10 Picnic at Wee Jasper while on a fossil hunt in 1971. From left to right: Professor Alfred Romer, David Brown, Mrs Romer, Mike Plane and Ken Campbell.
Photo: Gavin Young

Research

Several Visiting Fellows, some retired ANU staff and especially retired AGSO/GA personnel have had office accommodation and pursued their own research interests. Others have been short-term visitors giving occasional lectures and/or discussing research interests with colleagues. Many of the visitors were world authorities, giving our students and staff an exciting view of new developments. In the early years, we heard about palaeomagnetism from Professor Runcorn, transform faults from Tuzo Wilson, expanding Earth from Sam Carey, meteorites from Brian Mason, early atmosphere from Preston Cloud, vertebrate finds from Professor Alfred Romer (Plate 5.10), Scottish geology from Gordon Craig, the latest in sedimentology from Bob Folk, and granite studies from Leon Silver and E-an Zen, to mention just a few. Professor Lionel Weiss, a Visiting Fellow with RSES, initiated a series of popular tectonics seminars and field trips in 1965.

Over the past few years, the following have been actively working in the department for several months or more as Research Fellows or Visiting Fellows: Professor K. S. W. Campbell (vertebrate palaeontology), Professor W. D. L. Ride (vertebrate palaeontology), Dr R. E. Barwick (palaeontology), Dr D. Strusz (palaeontology), Dr G. Young (palaeontology), Dr P. Jones (palaeontology), Dr D. Lindley (palaeontology), Dr J. Caton (palaeontology), Dr E. Truswell (palynology), Professor S. R. Taylor (geochemistry), Dr K. A. W. Crook (sedimentology and marine geology), Dr A. Felton (sedimentology: rocky shores), Dr R. W. R. Rutland (Swedish Precambrian), Dr N. Exon (marine geology), Dr R. W. Johnson (volcanics), Professor G. M. Taylor (regolith), Dr W. Mayer (petrology), Dr R. V. Burne (sedimentology), Dr J. Magee (prehistory), Dr C. Klootwijk (palaeomagnetism), Dr A. Glikson (Precambrian and meteorites), Dr D. Mackenzie (geology of wine-growing areas), Dr R. Ahmad (sedimentology), Dr P. Hancock (economic geology). Reports of some of their individual work follow, and that of some Postdoctoral Fellows is included with reports of special projects in Chapter 6.

Professor W. D. L. (David) Ride

In December 1987, Professor W. D. L. (David) Ride retired from CCAE (now UC) and became Visiting Fellow in the Department of Geology, ANU. There, he continued previous studies of the fossil mammals of the Monaro and supervised PhD student Angela Davis in her study of the mammal fossil fauna of Yalcowinna near Cooma, and honours student Leanne Dancie (now Armand) in her study of the fossil mammals of Teapot Creek, Sherwood Station—also on the Monaro. Both studies were completed successfully.

David was awarded the Australian Centenary Medal 2003 by the Commonwealth Government for service in the field of palaeontology, and the Medal of the Riversleigh Society in recognition of his contribution to palaeontology. The latter was presented by Professor Michael Archer, the President of the Society, at its annual meeting at the Australian Museum, Sydney.

David was President of the International Commission on Zoological Nomenclature (ICZN) and Chairman of the editorial committee of the International Code of Zoological Nomenclature. On behalf of the International Union of Biological Sciences (IUBS) and ICZN, he attended meetings in Europe and the United States in preparation for the adoption of the fourth edition of this code and gave invited papers at the American Museum of Natural History, New York, and at the US National Museum, Washington, DC. In the field of Museum Studies, he was Chairman of the Executive of the Australian War Memorial and a member of the Council of the National Museum of Australia. He was active in the International Council of Museums (ICOM), participating in its organisation and program in Beijing.

As Vice-President of the IUBS, Paris, David was responsible for arranging to hold its international meeting in 1988 in Canberra. Then, on behalf of the IUBS, he was invited to review and report on the development of mariculture in Taiwan (two visits). He also participated in meetings of the IUBS in Hungary, Austria and Greece; in the last, he gave a paper on developments in zoological and microbiological nomenclature. David was invited by the Russian Academy of Science to give papers on the development of zoological nomenclature to the academy in August 1991 in Moscow and St Petersburg. During his time at the ANU, he has published or co-authored some 26 articles.

Richard Barwick

Richard Barwick came to the ANU when it was still the Canberra University College to teach zoology and remained an active member of what became the Department of Botany and Zoology (BOZO) until his retirement in 1994. His early research at the ANU, which eventually became a PhD, was on the physiology and ecology of lizards. After his retirement, Richard continued his collaboration with Ken Campbell on the study of fossil fish. Between them they have written more than 80 papers, all illustrated by Richard. These illustrations are works of art as well as meticulous recordings of anatomical detail (Plate 5.11). Richard's expertise and experience are always available to other members of the department, and he is involved in numerous projects covering a wide

Research

range of topics, including providing a large proportion of the illustrations for this volume. He has supervised large numbers of honours and postgraduate students over the years.

Richard Barwick's continuing passion is for the natural history of the Antarctic continent, which he first visited 50 years ago on the first New Zealand summer support party of the Transantarctic Expedition (NZTAE), in 1956–57. He was a member of the Victoria University (Wellington, New Zealand) Antarctic Expedition to the Dry Valleys of McMurdo Sound, and one of these is named in his honour. He returned to Antarctica in 2008 as a guest lecturer and guide on the tourist ship *Marita Svetlaeva*.

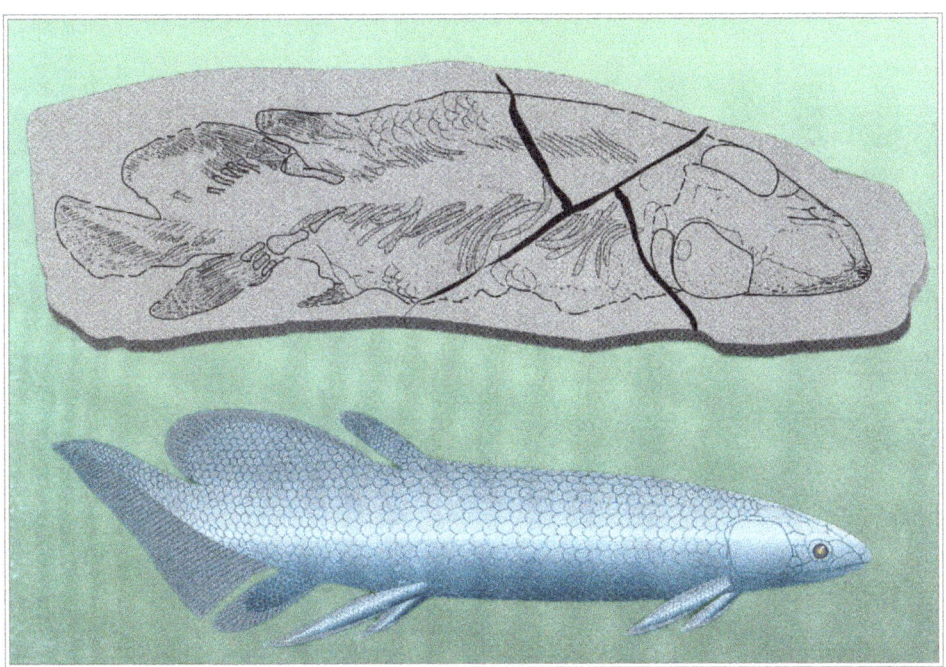

Plate 5.11 Fossil lungfish (Barwickia downunda). Found in the upper Devonian black shales at Mount Howitt, northern Victoria, described by Visiting Fellow Dr John Long.
Photo and drawing: R. E. Barwick

Judith Caton

Judith Caton, who undertook postgraduate studies in biological anthropology at the ANU, was jointly supervised by Dr Richard Barwick and Professor Colin Groves. She graduated in 1997, at the last ceremony marshalled by Mike Rickard. Judith came to the department early in 1997 to learn from Dr Barwick the techniques for illustrating her thesis, and remained there until 2007.

Geology at ANU (1959–2009)

Judith is primarily a zoologist, who is interested in functional morphology, ecophysiology and evolution of the gut in primates (Plate 5.12), including humans. These interests seem out of place in the Geology Department, but this was the one place in the ANU where others (namely, the geochemists) understood the use of heavy-metal tracers in the modelling of chemical reactors in nature, even if they knew little about 'the nastier parts of monkeys, apes and humans'. In 1998, Judith was awarded a visiting fellowship to continue her work on primate morphology at the Kyoto University Primate Research Institute in Inuyama, Japan. From 1999 to 2001, she was an ARC Postdoctoral Fellow, during which time she studied the role of fibre in the diet of humans and their modelling gut function. In her last year in the department, Judith produced a large proportion of the illustrations for the new edition of *Planetary Crusts* by Taylor and McLelland.

Plate 5.12 Judith Caton and subject at Perth Zoo

During her time in the department, Judith reinstituted the annual Open Week at the ANU's Kioloa field station and was appointed to the Kioloa Advisory Board, which helps with the management of the property. In 2007, she left the department as she was appointed the inaugural Visiting Fellow of the Kioloa Coastal Campus.

John Magee

Like Bob Musgrave's, John Magee's career illustrates the difficulties facing some dedicated and determined geological researchers. His career highlights are outlined in Chapter 11.

Desmond Strusz

Desmond Strusz is one of the longest-serving Visiting Fellows in the department. He began in 1997 after retiring from AGSO and continues his work to date. He is a world authority on Silurian brachiopods and has worked extensively on local fossils, especially from the Yass–Burrinjuck area. He has participated in five international conferences and has held or currently holds several important council positions, including: voting member of the International Subcommission on Silurian Stratigraphy of the International Union of Geological Sciences, until 2000, and, since then, corresponding member; Treasurer of the Association of Australasian Palaeontologists, 1996–97; Research Associate of the Australian Museum, from February 1996; Council of the International Working Group on Fossil Cnidaria and Porifera, at the Sendai Symposium, September 1999 until the Graz Symposium in August 2003.

Des assisted the department with tutoring and teaching of palaeontology to first and second-year students and on field excursions and mapping camps, especially to Delegate and Burrinjuck in 1998, to Burrinjuck and the Wellington–Burrandong area in 1999 and 2000, and to Gowan Green in 2000. He also assisted Tim Munson during 2000 in his work to prepare for publication the sections on Rugose corals from his honours and PhD theses.

Since joining the ANU, Des has published some 17 journal papers, mainly on Silurian brachiopods from the Yass area and nearby New South Wales. A paper on rhynchonellide brachiopods from the Silurian of Yass is almost complete, awaiting photography of Geoscience Australia specimens. His work on the spiriferide brachiopods from the Silurian of Yass will be the last of a series of papers on the Yass Syncline brachiopod faunas, based not only on earlier material held by the Australian Museum and the Geological Survey of New South Wales, but also collections made while at BMR/AGSO, and most significantly the collections made by the late K. J. Kemezys during the course of his PhD studies at the ANU. He has recently been involved in preparation of the *Guide to the Geology* of the Canberra Region, published in 2008.

Gavin Young

Professor of Geology Richard Arculus offered Gavin Young a Visiting Fellowship in the department in 1997. Research collections of fossil fish were moved from AGSO to a new laboratory area, and a close collaboration was resumed with Professor Ken Campbell and Dr Dick Barwick, who by that time were also Visiting Fellows in the department. Apart from a Visiting Professorship at the Museum National d'Histoire Naturelle, Paris (1999), and several periods in Germany (Museum für Naturkunde, Humboldt Universitat, Berlin) during 2000–04 as an awardee of the Alexander von Humboldt Foundation, Gavin has remained at the ANU, and is currently a Research Fellow in RSES funded by an ARC Discovery Grant (shared with Dr John Long, Museum Victoria). An international symposium on early vertebrates was organised as part of the International Palaeontological Congress held at Macquarie University in 2002, with a field excursion to many significant Devonian fish localities in south-eastern Australia, finishing at Wee Jasper and the ANU (Plate 5.1). This attracted extensive media coverage on local radio and television. Gavin edited the proceedings volume that was published in 2004. (Gavin's early history is reported in Chapter 11.)

Peter Pridmore

Peter Pridmore was with the Geology Department from July 1990 until the middle of 1997. Most of his work was undertaken as a Postdoctoral Fellow working with Professor Ken Campbell and Dr Richard Barwick. More than six publications on dipnoans, conodonts and marsupials resulted from this work. He also contributed to the history of the Australasian palaeontologists in *Rock Me Hard, Rock Me Soft: A history of the Geological Society of Australia.* Towards the end of his stay, however, when the research grants he was employed on had finished, Peter earned a small amount working as an editor for the Royal Zoological Society of South Australia, preparing their book *Wombats* (edited by R. T. Wells and P. A. Pridmore, Surrey Beatty & Sons), a compilation of 32 papers on various aspects of the biology of the three living species of wombats and their fossil relatives, which appeared in 1998, after he had left the ANU.

Peter Hancock

After 30 years as a professional geologist, consultant, company director and lecturer in the mining and engineering industries, Peter Hancock came to CRES at the ANU as a Visiting Fellow in 1991 to write a book, *Green & Gold: Sustaining mineral wealth, Australians and their environment* (ANU, 1993). This book, focused on sustainable and harmonious development of mining and community perceptions, had a salutary effect on the minerals industry and the more extravagant claims of the environmental movement. This led to research contracts for the Australian and Canadian governments as well as with industry, initially at the CRES and, since 1998, in the Geology Department at the ANU. His main contributions to the department have been through his links with industry and his role with the Australasian Institute of Mining and Metallurgy (AusIMM), mentoring students and assisting them to gain scholarships, company support, vacation work experience and providing industry careers advice. Peter's work is detailed in Chapter 9.

Professor S. R. Taylor

Professor S. R. Taylor (geochemistry) worked in the Research School of Earth Sciences (formerly the Department of Geophysics, RSPhysS) from 1961 until he retired in 1990, but had only brief contact with the Department of Geology during that time. In the 1960s, he collaborated with Allan White on work on andesites. About 1994, he gave a two-week course on planetology in the Geology Department in association with Warren Hamilton, who was visiting from the US Geological Survey and the Colorado School of Mines. From 1991 to 1999, Professor Taylor was a Visiting Fellow in the Department of Nuclear Physics, RSPhysS. Professor David Ellis then invited him to join the Deptartment of Geology as a Visiting Fellow, where he has been since 2000. Professor Taylor instituted a course in planetary science (now GEOL 3020), which he taught with Richard Arculus for four years, attracting about 40 students for each course. In 2001, he published a second edition of *Solar System Science: A new perspective* (Cambridge University Press), and he is currently finishing *Planetary Crusts: Their composition, origin and evolution* (Cambridge University Press), co-authored with Scott McLennan (State University New York, Stony Brook). During his tenure in Geology, Professor Taylor has mentored various students and has also published 14 papers in international scientific journals. In 2002, he was awarded the Walter H. Bucher Medal by the American Geophysical Union for 'original contributions to the basic knowledge of the Earth's crust', and, in 2008, he was awarded an Order of Australia.

Liz Truswell

Liz Truswell has been a Visiting Fellow since 1998. Her former position was as a Chief Research Scientist with the then Australian Geological Survey Organisation (AGSO). In that organisation, she was involved with leading an area of environmental geoscience, concerned with setting up programs focusing on climate change, on coastal zone issues, Antarctic geoscience, and a publicly oriented program on the geology of Australia's national parks. In the Department of Earth and Marine Science, Liz's research has focused mainly on the Antarctic, dealing with palynology, both as a dating tool (using it to date the onset of major glaciation on the Antarctic landmass, in conjunction with the Ocean Drilling Program) and as a monitor of vegetation and climate change in the Tertiary.

Research papers have dealt with pollen and spore assemblages from Prydz Bay in eastern Antarctica, which show that the late Eocene vegetation was a mosaic of dwarfed trees, scleromorphic shrubs and wetland herbs—analogous with the taiga found in the transition zone between the boreal conifer forest and tundra biomes across the Arctic Circle. Evidence points to the collapse of taller woody ecosystems during the Eocene–Oligocene transition and their replacement with tundra-like or fell-field vegetation during the Oligocene and Neogene. This temperature-forced regression seems to have been broadly synchronous across the continent. The high palaeolatitude location (~70°S) means that the Antarctic flora was adapted to long periods of winter darkness. Other work has dealt with the Miocene environments of Heard Island.

Science is, however, only one way of reconstructing the past. In 2000, Liz graduated with Honours in Painting from the Canberra School of Art. Now, working both as an artist and as a scientist, Liz finds, almost inevitably, that her geological background informs her art. One of her works, *Mesozoic margin— east Antarctica*, adorns the foyer of the Earth and Marine Science Building. This is an abstracted representation, in charcoal, of a Mesozoic forest, drawn over copies of one of her research papers on the same subject. Tearing up a reprint of one's own paper proved a wonderfully cathartic experience!

Plate 5.13 Stone Flowers: paintings of early fossil angiosperms by Liz Truswell
Photo: R. Barwick

In the past three years, Liz has held three solo exhibitions of her work. Two, *Drawing on the Past and The Stone Flowers* (Plate 5.13), were mounted in Canberra, and focused on fossil forms—the latter on the earliest angiosperms from the Cretaceous, which was the topic of her PhD thesis. In 2008, Liz held a major exhibition at the Goldfields Regional Gallery in her birthplace of Kalgoorlie, Western Australia. This again had the past as its focus, but here the view of the past was twofold: both the Earth's past as fossil forms and the imagery of a personal past growing up in a mining town. In all of her work, Liz uses the natural media of charcoal, in conjunction with the ochres of the Canberra region.

Working in both fields—science and art—has given Liz insight into just how closely related these two disciplines are. Both are motivated by the need to understand, and respond to, the natural world. This area—the interface between science and art—has become a burgeoning discipline in itself, and generates a lot of academic interest. Liz has given invited talks on this subject to a range of audiences—to the annual conference of the Australian Science Teachers' Association for one, and as part of Science Week activities in Canberra. A related issue is whether the sense of the aesthetic, which is in fact shared between science and art, can be seen to have some evolutionary significance. Questions such as: what part does this play in the evolution of the mind? Are we hard-wired for some kind of aesthetic response? Liz finds these questions intriguing, and she has presented conference papers reviewing the speculations related to these issues. She was elected a Fellow of the Australian Academy of Science in 1985, and, in 2007, was awarded a Fellowship of the Geological Society of Australia.

In summary, the position of Visiting Fellow in the Geology Department has provided Liz a base—both to remain in scientific research and to explore beyond that into the realms of the relationship between the arts and the sciences. Being able to work as a research scientist, a practising artist and a theoriser has given Liz enough challenges for a lifetime!

Peter Jones

Peter Jones, who obtained an MSc in the Department of Geology in the mid-1960s and a PhD (London) in the mid-1970s, returned to the department after he retired from AGSO. As a Visiting Fellow, he continued his research on bivalved microcrustaceans (Ostracoda, Conchostraca), Cambrian bivalved arthropods and Palaeozoic biostratigraphy. His peer-reviewed papers are on topics ranging from taxonomy, biostratigraphy and palaeo-biogeography to the biochronological correlation of the Devonian, Carboniferous and Permian time scales. As co-author, he has provided the biochronological input to papers on palaeo-biogeography, petroleum geology and magneto-stratigraphy. He has also written several consultant reports for oil-exploration companies in the Bonaparte Basin, north-western Australia. Peter is a corresponding member of the Subcommission on Carboniferous Stratigraphy of the International Stratigraphic Commission, and is one of a small team of international biostratigraphers invited by the GeoForschungsZentrum, Potsdam, to prepare a high-resolution correlation chart for the Devonian, Carboniferous and Permian systems. This chart was commissioned by German sedimentary geochemists in order to establish time control for the correlation of their stable-isotope profiles. Peter is grateful to successive heads of the Department of Geology (later Earth and Marine Science), and recently to the Director, RSES, for his appointment as a Visiting Fellow, which has allowed the necessary facilities and support for his research. During his time in the department, he has enjoyed his contact with students, and has greatly valued the friendship and support from staff members and former AGSO colleagues, including several 'retired' palaeontologists.

David Lindley

David Lindley was a graduate of the University of New South Wales and worked for 17 years in New Guinea. Having retired from that work, he bought a property at Good Hope, New South Wales, and continued with his first academic love of stratigraphy and palaeontology. He was made a Visiting Fellow in the department and began work on acanthodians from Taemas and Wee Jasper, publishing several papers on this topic. He has for a short term gone back to New Guinea, but hopes to return to the department to continue his palaeontological work.

Chris Klootwijk

Chris Klootwijk retired from Geoscience Australia (GA) in 2000, but continued his work at DEMS mainly on the carboniferous succession of the Tamworth Belt, southern New England Orogen, with the aim of better defining the late Palaeozoic pole path for Australia and to resolve a dispute with Phil Schmidt (CSIRO) and co-workers about the shape of the Australian and Gondwanan pole path. The ignimbrite-rich succession of the Tamworth Belt is particularly interesting because the ignimbrites retain a primary magnetisation component despite extensive magnetic overprinting. Chris has studied five tectonic blocks, as outlined by John Roberts: Rocky Creek, Werrie, Rouchel, Gresford and East and West Myall. Work on the Rocky Creek and Werrie Block has been published and writing up of the other blocks is under way. In the past few years, Chris has shifted fieldwork towards the Devonian of the Eastern Lachlan

Orogen, aiming to better define the middle Palaeozoic segment of the Australian pole path. Altogether, he has gathered plenty of interesting results and is now concentrating on writing the papers.

H. J. (Larry) Harrington

H. J. (Larry) Harrington assisted with tectonics lectures and attended seminars for several years. He has published studies of the tectonic relationships between New Zealand and eastern Australia. He ran a field excursion from Sydney to Canberra for participants in the Basement Tectonics Symposium, and acted as joint editor for the proceedings of the symposium. He also assisted with AusIMM industry presentations to students.

John Long

John Long (ANU Rothmans Fellow in Geology, 1983–84) left the ANU after winning a QEII Fellowship to go to work on the Gogo fish deposits at the University of Western Australia (1986–87). After this, he spent two years working on South-East Asian terranes and biogeography with Clive Burrett at the University of Tasmania (1988–89), before settling in to his job as Curator of Vertebrate Palaeontology at the Western Australian Museum (1989–2004). During this time, John was involved in many aspects of palaeontological research—from working the Gogo sites in the Kimberley to collecting dinosaurs and mega-fauna. He led three expeditions collecting complete marsupial lions in Nullarbor caves. His findings were published in *Nature* in 2007. His research focuses largely on the extraordinary three-dimensional fossils of Gogo fishes, based on his many expeditions to the site between 1986 and 2008.

His collaborations with ANU scientists Ken Campbell, Richard Barwick and Gavin Young continue to this day. His most famous recent discovery includes the world's oldest mother: a fish with an unborn embryo inside it, published in *Nature* (29 May 2008), for which he won the 2008 Australasian Science Prize. John has authored more than 200 scientific papers and popular articles and some 27 books, and in recent years has won national awards for his children's books on science and evolution. In 2001, he won a Eureka Prize for contributions to public knowledge, awarded by the Australian Museum, Sydney. In 2004, he took up a position as Head of Sciences at Museum Victoria in Melbourne. He has assisted US police in tracking down international fossil thieves and has also been a member of the brains trust for ABC Television's *Einstein Factor*. In 2009, he moved to the United States to take up a position as Vice-President of Research Collections at the Natural History Museum of Los Angeles County, California.

Geology at ANU (1959–2009)

Stefan Winkler-Nees

In *The Day After Tomorrow*, Roland Emmerich has made public what was long known to the science community—that global ocean circulation is one of the main drivers of Earth's climate. Still unknown is the time scale of its variability at which this circulation is having an effect on the human environment. That requires the work of marine geologists and palaeontologists, who need to recover marine sediments in order to reconstruct the history of oceanic changes and to try to work towards a scientific base for a predictive scenario.

This work has been done in some selected ocean basins—namely, the North Atlantic, for which Stefan Winkler-Nees (then Stefan Nees) was involved as a specialist in Quaternary benthic foraminifera at Kiel University. Stefan knew that his study region—the Norwegian–Greenland Sea—was perfect to investigate one of the key spots of the global circulation. He soon realised, however, that a real understanding of the system needs a link to the southern hemisphere, which allows a test of the validity of the 'North Atlantic theories'. And where would be better sites for comparison than in the seas south and east of Australia?

In 1991, Stefan wrote numerous letters to contacts in Australia, but most were answered ambiguously; no-one seemed to be really interested in his idea until he received a fax from Patrick De Deckker of the Geology Department, ANU. Stefan and Patrick soon started working on a proposal to the German Academic Exchange Service, which eventually brought Stefan to Canberra in 1992 for a few months as a Visiting Fellow. Samples were available from previous expeditions of US and German research vessels. But it soon became obvious that the material to work on was not sufficient to adequately provide the data needed. They started to prepare a grant application to the ARC for a follow-up visit for Stefan as a Postdoctoral Fellow to the ANU, which eventually allowed him to continue his work at the department for almost two more years. That period included a seagoing expedition on the *RV Franklin* with Patrick as chief scientist and an expedition with the French vessel *Marion Dufresne* to the Indian Ocean. In order to continue working on material from both cruises, Stefan applied successfully for a postdoctoral grant from the German Research Foundation, which included a number of visits to the ANU Geology Department in the 1990s. He also used this time to prepare, together with Patrick, an application to bring the German vessel *Sonne* and the *Marion Dufresne* to their study region: the Tasman Sea and the Australian and New Zealand sectors of the Southern Ocean. These cruises eventually took place in 1996 and 1998 and brought together a scientific crew from Australia, France, Germany and New Zealand. In 2000, the US drill ship

JOIDES Resolution visited the sea south of Tasmania and brought to the surface material documenting the opening of the 'Tasman Gateway' and the onset of the global ocean circulation.

Material from all these cruises was analysed at many places in the world and results were published in major journals. This demonstrates the international character of a scientific idea, which grew and evolved successfully in the Geology Department. Stefan's first visit to the Geology Department at the ANU had started a remarkable collaboration between Patrick, his students and numerous colleagues worldwide, and Stefan is happy to have been part of this story. Since 2007, he has been working at the German Research Foundation (Bonn, Germany), responsible today for establishing an open-access policy and infrastructure for research-relevant raw and primary data.

Charlotte M. Allen

Charlotte M. Allen arrived in Canberra in May 1991 from Kentucky, USA, and worked in the department until June 1998 in three postdoctoral positions, with Bruce Chappell and with David Ellis. She arrived just a few months before the move to the current building. The personnel who influenced her most during the first few years (aside from Bruce) were Phil Blevin, a co-Chappell Postdoctoral Fellow, David Champion, who was finishing his PhD, and Penny King, who was doing her honours. During those first couple of years, Charlotte undertook fieldwork in Queensland between Townsville and Mackay, trying to characterise and date the granites there. Both David and Penny were field assistants. The last few years of Charlotte's Postdoctoral Fellowships were spent doing experiments in the Ca-Ti-Zr system with David Ellis, with a view to characterising some of the minerals used in radioactive waste storage.

If Charlotte had not insisted that her PhD advisor send her in 1987 to the first Hutton Conference in Edinburgh, with a field trip in Donegal, where she met Bruce Chappell, Allan White and Ian Williams, she would not have been offered the postdoctoral post with Bruce, would not have met her husband (who worked at the ANU), and would not be living out her days in Australia. Charlotte has worked continuously for the ANU since her arrival in Australia; she is currently a Technical Officer in RSES in charge of running a quadrupole ICP-MS.

Geology at ANU (1959–2009)

Plate 5.14 Éva Papp in Antarctica
Photo; from É. Papp

Éva Papp

Éva Papp graduated from Loránd Eötvös University, Hungary, in 1982. She arrived in Australia with her husband and baby daughter in 1989. After a few years at Monash University as a Research Assistant, she won an Australian Postgraduate Research Award and completed a PhD in Geophysics at the University of Western Australia while working part-time in the CSIRO Mineral Mapping Technologies group. She arrived at the ANU in 1998 as a CRC-LEME-appointed Research Fellow, based at AGSO, working on advancing shallow geophysical techniques for regolith exploration. She also supervised honours students and contributed to lecturing in the Geology Department in Prame Chopra's geophysics courses. For the summer of 1999–2000, she had the opportunity to participate in the fifty-third Australian National Antarctic Research Expedition (ANARE) at Casey Station (Plate 5.14). Currently, she is working as a consultant based in Canberra, mainly in groundwater exploration. She continued to be involved in casual teaching of geophysics in the department until 2006.

Dirk Marten Kirste

Dirk Marten Kirste was a researcher in hydrogeology and hydro-geochemistry at Geoscience Australia between 2000 and 2003, when he joined the department as a Research Fellow as part of the CRC-LEME. He resigned in 2006 to take a faculty position at Simon Fraser University in Canada. While at the ANU, his research and teaching focused on salinity, acid-sulfate soils, hydrogeology, hydro-geochemistry and regolith geoscience.

Susan A. Welch

Susan A. Welch graduated from the University of Delaware with a PhD in 1997. She was then a Post-doctoral Fellow and assistant scientist at the University of Wisconsin until 2003, when she joined the department as a Research Fellow as part of the CRC-LEME. She left in 2008 at the end of her contract to take a position as a Research Scientist at Ohio State University. While at the ANU, her research projects included the study of acid-sulfate soils, bio-geochemistry and the role of microbes in mineral weathering and element mobility, salinity dynamics and regolith geoscience. From 2004 to 2006, Susan lectured in geochemistry.

Harvey Marchant

Harvey Marchant was a Visiting Fellow in the department from 2005 to 2008, after his retirement from the Antarctic Division, where he was head of Australia's Biology Research Program. He was also a Professorial Fellow of the Department of Zoology and the School of Botany at Melbourne University. His main research interests are the interactions between aquatic micro-organisms, and he has visited Antarctica 15 times as part of Australian, Japanese and American research programs.

He is the author of more than 150 scientific publications and book chapters, as well as numerous contributions to popular magazines. He was the senior editor of the book *Australian Antarctic Science: The first 50 years of ANARE* (2002), and co-editor of *Antarctic Marine Protists* (2005). While a Visiting Fellow, Harvey was a co-author of *Antarctic Fishes* (2006), was on the Australian Research Council's Group of Experts, and remains on the Research College Board of the University of Tasmania. Harvey was a lead author of the Intergovernmental Panel on Climate Change scientific assessments (1995, 2001 and 2007) for which he was a joint recipient of the 2007 Nobel Peace Prize. He has contributed to undergraduate lecture programs and supervision of postgraduate students and remains involved in the national program Scientists in Schools. Harvey represented Australia on the Scientific Committee on Antarctic Research on the Life Sciences and was a member of the steering committee of the Ecology of the Antarctic Ice Zone program.

Other long-term collaborators who spent several weeks in the department, or made repeat visits, were Professor S. Banno (metamorphic petrology), Professor A. J. R. White (petrology), Dr W. Yim (surficial), Professor S. Guggenheim (mineralogy), Professor L. Silver (petrology), Professor E. H. T. Whitten (statistical petrology), Dr E. Zen (petrology), Professor Y. Zheng (structural geology) and Professor Fin Campbell (economic geology). Several former ANU staff also became long-term Visiting Fellows, including Professor Brown, Professor Campbell, Professor Eggleton and Professor White and Dr Crook and Dr Rickard.

Apart from those mentioned in personal accounts of departmental members, other publications by Visiting Fellows include the Chinese translation of Mike Rickard's booklet on *Geological Map Interpretation* by Yadong Zheng (1981), Wolf Mayer's booklet in 1996 on *The Building Stones of Parliament House*, and Professor Richard Stanton's book on economic geology. E-an Zen published an interpretation of Lachlan granites, and Tim Whitten published a statistical review of the geochemical features of Lachlan granites.

Postgraduate Research

(Graduate students and theses titles are listed in Appendices 3, 4 and 5.)

Until recently, most of our research students came from overseas, because the department insisted on not taking our own graduates for higher degrees. Several of our honours students transferred to RSES for graduate studies, however, and because of this policy we did not have the altercations over pirating of potential graduate students that some other ANU departments had. Unfortunately, other universities did not follow our lead in enforcing student mobility. Eventually, this policy was dropped, especially when we were the only university to offer some specialist studies—for example, regolith. In addition, government-imposed fees for foreign students severely affected the number of overseas applicants, as it was impossible for the department to defray the cost of foreign student visa fees. Nevertheless, graduate students have been attracted from many places, including Belgium, Canada, China, Colombia, Czechoslovakia, France, Germany, Indonesia, New Zealand, Sierra Leone, Sweden, Papua New Guinea, Russia, the United Kingdom, the United States and Vanuatu, and they have worked in numerous projects all over Australia, some with exploration companies and others with BMR/AGSO/GA field projects—for example, to central Australia and Mount Isa.

Staff retirements caused problems with the supervision of some graduate students, especially those working in very specialised fields, and, as a result, some scholarship extensions were sought and obtained. Thanks are due to Professor John Chappell (RSPacS, later RSES), Russell Korsch, P. O'Brien and

A. Wells (AGSO), Brian Jones (University of Wollongong), G. Brierley (Macquarie University) and G. Taylor and K. McQueen (UC) for taking over supervision duties at critical times.

One of the earliest students (1963) was **Hans Steiner** from Alberta. He mapped out the stratigraphy and detailed sedimentology of the upper Devonian Merrimbula sequence. During this project, Keith Crook arranged for a reconnaissance flight along the coast but was unable to come himself, so Henry Charlesworth (a Visiting Fellow from Canada), Mike Rickard and Hans—three burly, bearded blokes—climbed into the very light plane to be piloted by a 'little slip' of a girl at very low level down to Green Cape and back: safely! Later, Hans was granted an extension of his scholarship for a year to work on his theory on the relationship between geological periodicities and galactic rotation. This controversial and remarkable work was eventually published by the GSA in 1967.

Others among the earliest research student cohort were **Mel Stauffer** (also a Canadian), who expounded the nappe structure of the local Ordovician greywackes—the first to be reported in Australia; **Dennis Belford**, our first Ampol Scholar, who worked on micro-fossils; **Neil Powell**, who worked on bryozoans and, when applying for a job in Ethiopia, was offered lectureships in both zoology and geology. **Gerry Reinson**, another Canadian, mapped swamp deposits at Mallacoota in Victoria. **Jeff Harris** worked on the geochemistry of mineral deposits in New South Wales; and **Mel Abbott** started work on the (New England) granites with Bruce Chappell. Other early petrology students were **Mike Rhodes**, who worked on the granites of the Hartley district for his PhD in 1969, then spent a year as a postdoctoral student at Northwestern University, before going to the Manned Spacecraft Center in Houston, where he was responsible for operation of the XRF instrument that analysed lunar samples. This developed into a broader interest in the compositions of basalts, which he has followed up as a professor at the University of Massachusetts, Amherst. He has been heavily involved in research into sea-floor basalts and the Hawaiian volcanoes. **Doug Haynes** studied the geochemistry of altered basalts and associated copper deposits with Allan White and Bruce Chappell for his PhD in 1972, before joining Western Mining, where he was instrumental in the discovery of the Olympic Dam deposit.

Most staff have spent time in the field supervising graduate student work all over Australia and New Guinea, some in association with BMR/AGSO field parties. Joe Morrow and John Vickers accompanied and supported several students as drivers and camp managers.

Geology at ANU (1959–2009)

Peter Cook worked on the southern nappe front west of Alice Springs and, together with Keith Crook, published alternative explanations for the circular Gosse Bluff structure. **Brian Jones** mapped the stratigraphy of the northern Amadeus Basin. **Roger Majoribanks** set up a luxury fly camp at Ormiston Gorge to map the Proterozoic nappe structures. **John Funk** worked out of the BMR Anbalindon Camp on the Heavitree quartzite, and **Donald Windrum** mapped the high-grade metamorphics in the Strangways Ranges. **Ramon Looseveldt** mapped around Cloncurry and **Priharjo Sanyoto** in the western Isa belt from the AGSO Mount Isa field camps (see Chapter 11 for more details).

The Yass–Taemas area of New South Wales is well exposed and contains abundant, remarkably well-preserved Silurian and Devonian fossils. This area became a major focus of palaeontological research for the department under Professor Campbell's supervision. Towards the northern end of the Yass Basin, some Silurian sections are continuously exposed and the late **Kazys Kemezys** undertook a study of the brachiopods from the Black Bog shale for his PhD degree. He published some short theoretical papers, but did not complete the systematic work (see comments by Des Strusz in the previous section). **Brian Chatterton**, a graduate from Dublin, started work on the brachiopods from the Receptaculites and Warroo limestones at Taemas. These were silicified fossils in limestone, and, by etching, they produced specimens of such high quality that their publication revolutionised the study of their groups. In addition, whole series of ontogenetic (early growth) stages of trilobites appeared. Some of these were completely new. Brian Chatterton was kept on as a Demonstrator while this work was published; it is now a major part of the new volume of the *Treatise on Invertebrate Paleontology*. **Paul Johnson** studied the pelecypods of the Bloomfield area, before returning to Canada to the Tyrell Museum in Alberta (see Chapter 11). **Albie Link** made detailed studies of the conodonts and stratigraphy of the Mount Bowning area south of Yass. This work later contributed to the dating of the Silurian–Devonian boundary by radiometric work at RSES. **Linda Shields** from Aberdeen, Scotland, worked on the ostracods from the Taemas limestones. She presented the work as an MSc but it should have been continued for a PhD. It was highly commended and published in *Palaeontographica*. **David Feary** made detailed analyses of the Silurian limestones in the Boambolo area, using quantities of acid to clean rock faces to expose fossils and sedimentary structures.

In other areas, **Peter Jell**, from Queensland, undertook the study of pagetiid trilobites for a PhD. He had already done fieldwork in the Georgina Basin before he came to Canberra, and he gained access to specimens from China, America and Europe. He also undertook a study of the global distribution of Cambrian trilobites. His pagetiid work was contributed to the international *Treatise on Invertebrate Paleontology*. He was awarded an ANU PhD Scholarship and spent a year in New York. **Robert Day**, also from Queensland, had much experience in the Cretaceous rocks of the Artesian Basin (see Chapter 11). Following the work of Whitehouse, he had collected a large number of bivalves and ammonites from the Roma and Tambo series. The department employed him as a Demonstrator and as a PhD student to work on these specimens. This work was of value not only for the outcropping specimens, but for the buried material of the basin.

The department and its undergraduate students had examined the Permian marine sediments of the NSW South Coast for a long period, collecting many specimens of crinoids preserved as limonitic replacements or as casts. **Robert Willink** from Tasmania undertook to study these faunas because the palaeolatitude of the sites was approximately 60° south, and most of the described forms had come previously from the warm waters of the northern hemisphere. The faunas were just as exciting as expected, and displayed a number of features unknown elsewhere, as well as groups of an intermediate kind. This work, in thesis form, was completed just in time to be included in the international treatise. In 1992, **Angela Davis**, while working on vertebrates in Quaternary lake-bed sediments, made international press with her discovery of a 7000-year-old Aboriginal burial site near Cooma.

A large group of honours and graduate students has worked on micropalaeontolgical projects with Patrick De Deckker and numerous Visiting Fellows (see the section on Marine Research in Chapter 6). Several students have worked on local stratigraphy, sedimentology and structure. **Victor Gostin** studied the Permian fluvioglacial sediments at the southern end of the Permian outcrop near Ulladulla. **Peter Hayden** made a detailed structural analysis of the Ordovician rocks of the Cullerin Horst north of Cooma. Numerous honours students worked in the Tumut area with Keith Crook, which culminated in a PhD study by **Peter Stuart-Smith**. A Faculty Research Grant enabled Crook, Rickard and Stuart-Smith to undertake the first seismic traverse in the Lachlan Fold Belt across the Tumut Trough. This work was carried out by AGSO, supervised by Dr Jim Leven, who used ANU students as 'juggies' to run out the recording lines. The project terminated prematurely as the team ran out of dynamite. The results of this labour were published in *Tectonophysics* in 1992.

From 1983 to 1994, field-based projects (mapping plus some study of mineralisation) were undertaken by more than 20 honours students who wanted to work in the mining/exploration industry. Students working with

John Walshe focused on the metallogeny of the Tasman Fold Belt. Pragmatically, a good deal of the work was done on mines and prospects within four hours' drive of Canberra. Studies on some of the more significant deposits were undertaken by postgraduate students **Paul Heithersay** (Cu-Au at Goonumbla, North Parkes), **Ren Shuang-Kui** (Ardlethan Sn), **Scott Halley** (Mount Bischoff Sn, in western Tasmania), **Bill McKay** (Woodlawn Cu-Pb-Zn), **Idunn Kjolle** (Browns Creek Au), **Eugeniy Bastrakov** (Lake Cowal Cu-Au) and **Bill Platts** (base-metal, gold and barite mineralisation in the Canberra region). Not all the effort was focused on the Tasman Fold Belt. Other activities included PhD studies by **Roger Skirrow** on the Tennant Creek Au-Cu-Bi deposits and **Greg Cameron**, who studied the Porgera Au deposit, Papua New Guinea, as well as an MSc study by **Imants Kavalieris** on the Gunung Pani Au prospect in northern Sulawesi.

Three ARC-funded Postdoctoral Fellows worked with John Walshe. **Dr Stewart Eldridge**, a graduate of Pennsylvania State University, was employed from 1984 to make the first in-situ measurements of sulfur isotopes on the ion probe at the Research School of Earth Sciences (RSES). In the following decade, he applied this technique to a diverse range of geological problems, from the Pb-Zn deposits of the Australian Proterozoic to the sulfur isotopic variation of sulfides in geothermal systems and the sulfur isotopic composition of sulfide inclusions in diamonds. He joined Western Mining Company about 1994 and worked at Olympic Dam for a time as the Chief Geologist. His work on in-situ stable isotope analysis was a genuine pioneering effort. After a break of more than a decade and following technological advances, RSES is once again establishing the capacity to make in-situ stable isotope studies. At the time he left the ANU, Stewart was the longest-serving visitor at the RSES.

Dr Caroline Perkins' research, funded through an ARC Linkage Grant and Australian Mineral Industry Research Association (AMIRA) project (P334A), 1990–96, made the first systematic age determinations of mineral deposits in the Tasman Fold Belt system using Ar-Ar dating techniques. She worked jointly with the Ar-Ar laboratory at the RSES and worked and published extensively on a diverse range of problems, including the geochronology of the Mount Read volcanics; timing of shoshonitic magmatism and mineralisation in central-western New South Wales; and Ar-Ar geochronology on the Mount Isa Cu ore body. **Dr Steve Lonker** was an ARC-funded Postdoctoral Fellow, who worked on mineral chemistry of geothermal systems in New Zealand and Iceland from 1989 to 1992. Other research projects are outlined in Chapter 6.

6
Major Research Projects

The Geology Department has been very successful in obtaining research grants, large equipment grants and grants of ship time from international ventures. Full listing of these would be too lengthy, but, for example, more than $1.8 million was awarded in 1995. Hundreds of research papers and books have been published. In 1988, it was reported that over the 30-year life of the department, staff and associates had published at a rate of 1.4–4 (average 2.7) publications per staff member per annum, and the citation rate had increased from 104 to 294 over the period 1983–87. In 2005, 57 journal articles were published, and 89 in 2006. In 1995–2005, ANU (DEMS and RSES combined) rated eighth in the world for citations by Geoscience Universities, with an average 15.6 citations per paper. Recent publications can be viewed on the department's web site (<http://ems.anu.edu.au>). Five of the department's major research projects are described below.

Geochemical Evolution and Metallurgy of Continents (GEMOC)

Australian Research Council (ARC) funding for a National Centre for Geochemical Evolution and Metallurgy of Continents (GEMOC) was awarded in January 1996 to the School of Earth Sciences at Macquarie University and the Department of Geology at the ANU, in collaboration with CSIRO Exploration and Mining and AGSO (Plate 6.1). Professor Suzanne O'Reilly (Macquarie University) was director and Professor Richard Arculus co-director and coordinator of the research and teaching program at the ANU. Dr David Ellis acted as co-director whilst Richard Arculus was on study leave.

Integrated studies were to be carried out at all scales from micron size to whole-Earth processes, interfaced across the boundary of geology and geophysics with cooperation from industry. A new energy dispersive XRF (spectrometer) unit was acquired by the ANU. Dr Prame Chopra was appointed to teach remote sensing and train staff and students in the use of a 'Map Info System'. The rest of his time was allocated to the 'Hot Rock' project with

Dr Doone Wyborn. Others associated as Visiting Fellows were Dr Jan Knutson, Professor P. W. Candela, Dr R. W. Johnson, Dr I. Metcalfe, Dr S. Okamura, Dr J. Sheraton, Dr R. G. Warren and Dr C. Allen. Charlotte Allen worked with David Ellis on the stability of titanite in the crust, and she also worked on the partial melting of eclogite. Dr Phillip Blevin, Dr Stephen Eggins, Dr Ian Parkinson and Dr Tezer Eszat were working on geochemical analyses. Professor Bruce Chappell worked on the granites of eastern Australia, and he made available his extensive database of chemical analyses of granites and related rocks for relevant projects with GEMOC. Dr John Walshe supervised students working on mineral deposits. Professor Arculus worked on the genesis and tectonics of Western Pacific island arcs, and supervised Carl Spandler in honours and PhD studies of 'the roots of the arc' using the Greenhills ultramafic complex in New Zealand. Other students associated with GEMOC projects were Ulrike Troitzsch, Tony Kemp, Colleen Bryant, Greg Cameron and Eugene Bastrakov (PhD students) and honours students Margaret Chorley, Katie Devlin, Megan James and Ian McIntosh.

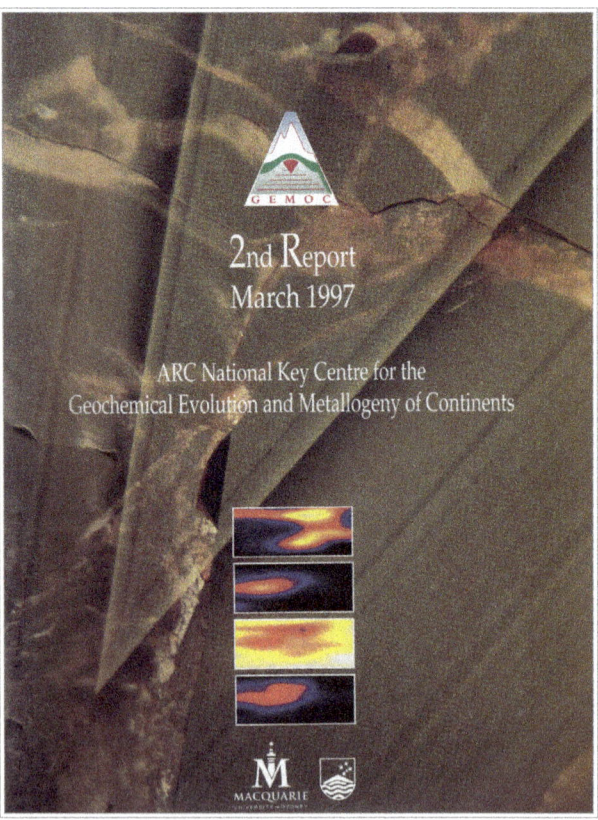

Plate 6.1 Cover of GEMOC second report

Australian Mining Industry Research Association (AMIRA)

Phil Blevin arrived at ANU Geology in 1989 after completing a PhD at James Cook University of North Queensland (JCUNQ). He commenced postdoctoral studies with Bruce Chappell as part of Bruce's industry-funded Australian Mining Industry Research Association (AMIRA) grant on a series of projects titled 'Geochemistry of granites as an aid to mineral exploration'. This long series of successful projects undertaken by Blevin and Chappell (the P147 series with two extensions) ended in 1994.

Not long after Phil's arrival in the department, Bruce 'volunteered' him for the post of secretary for the second International Hutton Conference on granites and related rocks, which was held at the Australian Academy of Science Shine Dome in Canberra in September 1991. The conference went off well despite a blizzard interrupting the progress of the pre-conference field trip through the Snowy Mountains, and Phil having a frantic search around the campus after he arrived at the Audiovisual Unit to pick up four projectors and 20 slide carousels only to find there was no record of the original request. One outcome of the conference was the return of Professor Philip Candela (University of Maryland) to the Geology Department in 1994–95 to work with Phil on unusual textures in granites, which has subsequently entered the geological lexicon as 'interconnected miarolitic textures'.

AMIRA Project 147 was followed in 1994–97 by a major collaborative ARC–AMIRA project (AMIRA P425) involving 17 industry and government sponsors. Gregg Morrison undertook a three-year study of the magmatic and hydrothermal evolution of intrusion-related copper and gold deposits in eastern Australia as part of this project. Greg came from Klondyke Exploration Services, with whom he had worked on gold deposits in northern Queensland, as well as on other AMIRA projects. His experience was vital in helping to extend the granite geochemistry to a more applied deposit focus. Deposits and related local and regional studies were undertaken in New South Wales (Cadia, North Parkes, Copper Hill and Temora) and Queensland (Kidston, Mount Leyshon, Ravenswood and Red Dome). AMIRA described it as one of the most successful projects of its type up to that time. The final AMIRA project (P515: Igneous metallogenic systems of eastern Australia) ran from 1999 to 2002 and produced a first-pass synthesis of the granite geology and related metallogeny of eastern Australia. Dr Caroline Perkins also worked under AMIRA Grant P334A dating mineral deposits.

Geology at ANU (1959–2009)

During his time at the ANU, Phil Blevin supervised a steady stream of honours students: Anthony Budd (1993), Nicola Harvey (1994; Masters in 2000), Greg Kovac (1996), Margaret Chorley (1997), Ben Cairns (1997), Margaret Spandler (1998), Natalie Kositcin (1999), Damien Dempsey (1999), Georgie Burch (1999), Anthony Johnston (2000) and Dan Isaacs (2000). Many of the students were able to undertake SHRIMP dating as part of their honours projects through a special arrangement with Dr Ian Williams of the RSES.

Phil subsequently became a Visiting Fellow in the department, leaving in 2003 to work as an industry consultant. In late 2005, he moved to Port Stephens and commenced work with the Geological Survey of New South Wales (which had recently been relocated from Sydney to Maitland), initially as a senior geoscientist responsible for strategic assessments, and currently as a research scientist and leader of the mineral-systems team.

Plate 6.2 CRC-LEME poster
From Bear McPhail

Major Research Projects

Cooperative Research Centre for Landscape Evolution and Mineral Exploration (CRC-LEME)

In 1995, the Cooperative Research Centre for Landscape Evolution and Mineral Exploration (CRC-LEME) was awarded jointly to the CSIRO Division of Exploration and Mining (Perth), Geoscience Australia Regolith Group and the Centre for Australian Regolith Studies at ANU and UC (Plate 6.2). This provided a great opportunity to increase both staff and postgraduate student research activity in regolith geology (Plate 6.3). Berlinda Crowther was appointed to support field and laboratory work and Judy Papps managed the LEME office. David Tilley took over some undergraduate teaching opportunities, leaving Tony Eggleton free to manage the ANU node.

Plate 6.3 First-year students examine weathered granite at Island Bend in the Snowy Mountains, New South Wales
Photo: T. Correge, from the 1988 Annual Report

In 2001, CRC-LEME received renewed funding. The slight name change at this time to the CRC for Landscape Environments and Mineral Exploration reflected growing involvement in environmental science. On Tony Eggleton's retirement, Ken McQueen took over as the LEME Assistant Director for Canberra. Bear McPhail was the ANU representative for LEME and was on the executive, as well as being a LEME key researcher. Ian Roach was appointed MTEC Regolith Geoscience Lecturer, and later Education Program Leader. Susan Welch and Dirk Kirste were appointed as Research Fellows in 2003 and Matthew Lenahan was appointed as Research Fellow in 2007, following the completion of his PhD with LEME. Many other members of the department, as well as researchers from other parts of the ANU, were affiliated with CRC-LEME. During the second incarnation of LEME, 27 honours students and 15 PhD students from the department were supported by CRC funding.

Major research initiatives within the department that were made possible by funding from the CRC included studies of groundwater geochemistry, biogeochemistry, the history of aridity and landscape evolution, metal mobility, salinity, acid-sulfate soils, mineral weathering and geochemical exploration (Plate 6.2). Much of this work has been published in CRC-LEME reports and conference volumes, plus a growing number of journal articles and book chapters. Publications from this research will continue into the future. Examples of published topics include salinity dynamics in the Barmedman area of central New South Wales and alkaline soils in the Bellata area of northern New South Wales; the impact of bacteria on gold mobility and mineral weathering; understanding acid-sulfate soil systems in evaporation basins of the Murray River, South Australia and Kempsey, New South Wales; dune formation and transport in central Australia; experimental studies of copper and zinc sorption and transport; geochemical signatures in regolith and groundwater in the Lachlan Fold Belt; and the character and origin of the Weipa bauxite.

In addition to many journal articles, extended conference abstracts, CRC-LEME open file and restricted file reports, several books and book chapters have been published on regolith geoscience by department researchers, including *The Regolith Glossary* (edited by R. A. Eggleton, 2001, CSIRO), *Regolith Geology and Geomorphology* (Taylor and Eggleton, 2001, John Wiley & Sons), and *Regolith Science* (edited by Scott and Pain, 2008, CSIRO), as well as a special issue of the *Australian Journal of Earth Sciences* on the Weipa bauxite (edited by Eggleton and Taylor, 2008).

Major Research Projects

LEME Student Research

Wang Qi-Ming, one of the department's first students from China, investigated the alteration by weathering of pyroxenes and amphiboles in the Mount Dromedary complex of south-eastern New South Wales. On the basis of Wang's electron microscope work, Jill Banfield included the alteration of biotite in her honours thesis on granites of the Bega Batholith, and then extended her studies to a masters thesis by looking at granite weathering in detail, and in so doing explained how plagioclase, microcline and biotite weathered. Mehrooz Aspandiar, who came from India to do a Graduate Diploma, worked on chlorite weathering from altered basalts of the NSW South Coast for his masters thesis, before undertaking an innovative study of the regolith in the Charters Towers region of Queensland for his PhD.

David Tilley began his time in the department as a PhD student investigating the origin of pisoliths in the Weipa bauxite. David stayed on as an Associate Lecturer, taking over many of Tony Eggleton's teaching duties when CRC-LEME was established. Leah Moore, from New Zealand, travelled the length of the Australian east coast in her field trips to collect basalt weathering profiles. Her work showed that the processes involved were essentially independent of climate or age. Another student from China—in this case, via New Zealand—was Ma Chi, whose PhD work on the ultra-structure of kaolinite gave new insights into how this mineral acts in the regolith and soil to retain transient ions through the presence of single atomic layers of smectite on each crystal's outer surface. Watcharaporn Keankeo from Thailand—also via New Zealand—worked on garnet alteration under both high-temperature (eclogite intrusion) and low-temperature (weathering) conditions.

Six more PhD students, benefiting from CRC-LEME funding and the wider range of guidance available from its scientists, undertook regolith research based on broader geological themes than the earlier mineralogical focus. K.-P. Tan worked on understanding the Portia Prospect—a project limited entirely to drill-core material, as all the rocks of interest lay deep beneath the plains between Broken Hill and the Olary Ranges. Steve Hill broke new ground as he mapped and interpreted the regolith history of the Broken Hill region. His sister, Leanne, was also a pathfinder—in her case investigating the relation between the regolith and the plant communities growing in it. Eric Tonui had the unusual opportunity of making detailed analysis of the regolith at a new open-cut mine developed at Goonumbla, near Parkes, in New South Wales. Juan-Pablo Bernal, a scholarship student from Mexico, tackled the difficult problem of dating regolith events. Using U-series geochemistry, he established ages in the order of 100 000 years for pisoliths in the upper regolith of the Ranger uranium mine in the Northern Territory. Taking advantage of these projects based on the long-term weathering of undisturbed rocks, Greg Shirtliff chose to work on the weathering of mining waste-rock piles at the Ranger mine, in an effort to understand how they react to rain water and what might be the implications for their long-term stability.

In LEME 2, there were many PhD studies, with many reflecting the stronger focus on environmental research in the regolith. These included Frank Reith (the effects of microbiota on gold mobility in the Australian regolith), Alistair Usher (aqueous gold chemistry), Chris Gunton (the effects of acidity and salinity on the sorption of copper and zinc on goethite), Matthew Lenahan (the hydro-geochemistry of a saline aquifer system), Michael Smith (the mineralogy and hydro-geochemistry of alkaline, saline-soil scalds), Kathryn Fitzsimmons (the geochronology, evolution and transport of dunes in central Australia), Kamal Khider (the regional geochemical dispersion of elements in the regolith of the Girilambone region of central-western New South Wales) and Luke Wallace (geochemical and mineralogical characteristics of the Loveday Basin, an inland acid-sulfate soil environment).

Hot-Fractured-Rock Project

This project was set up at the ANU (as Hot-Dry-Rock) in 1997 and was funded by an Australian Greenhouse Office Renewable Energy Commercialisation Program grant. Dr Prame Chopra and Dr Doone Wyborn moved from AGSO to carry out this work. The hot-dry-rock concept involves drilling into buried granite and pumping cold water down and hot water back up to drive turbines to produce electricity. Initial research partly carried out at AGSO set out to locate areas in Australia where hot granites were blanketed by thick insulating sediments (Plate 6.4). A crucial aspect was to determine the local stress fields, as several overseas projects had failed because stresses in the granites were vertical and thus prevented horizontal flow of water. Their research showed that the maximum principal stresses were favourably horizontal in the areas of interest in the Hunter Valley of New South Wales and remote areas of South Australia. Initially, a 2 km well was drilled in the Hunter Valley, but the focus then moved to South Australia. This exciting project to produce clean power has now gone commercial with a pilot plant under development extracting heat from granite beneath the Cooper Basin in north-eastern South Australia.

Major Research Projects

There is now a rush to commercialise these projects, and Australia is close to leading the world in the field. Doone and Prame were founding directors of the publicly listed company Geodynamics Limited that spun out of the ANU research. The company is based in Queensland and is now one of the top companies from Queensland listed on the Australian Stock Exchange (ASX), with a market capitalisation of about $300 million. The company's success in initiating the development of Australia's potentially vast geothermal energy resources has led to many other companies taking up the challenge. There are now something like 27 companies involved in geothermal research, exploration and development in Australia. Six of those are listed on the ASX. The proposed exploration budgets across Australia currently stand at about $700 million. This new industry has sprung up in the past few years on the back of Geodynamics' results.

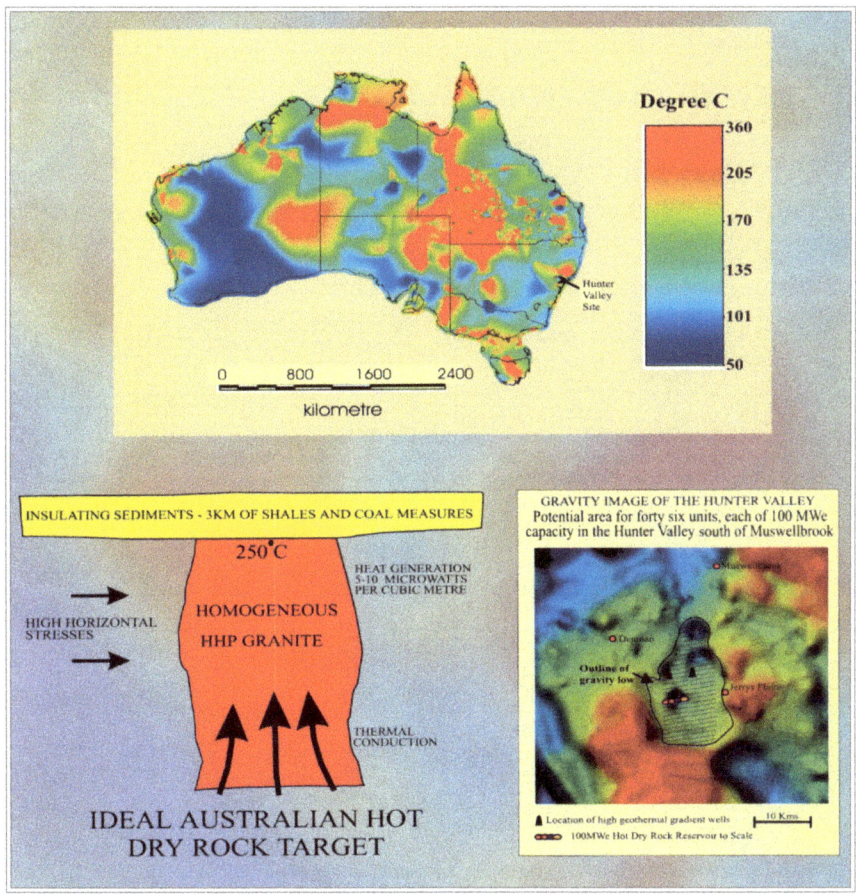

Plate 6.4 The Hot-Fractured-Rock Project explained
From the cover of the 1996 Annual Report, from a poster by P. Chopra and D. Wyborn

119

In its own work, Geodynamics has drilled into granitic rock at a depth of more than 4 km where the rock temperature is 250°C. The location near Innamincka in South Australia is thought to contain the hottest rocks in the world at this depth outside volcanic centres. Within the rock mass, Geodynamics has created the world's largest artificially enhanced underground heat exchanger (or reservoir). In 2005, water and steam flowed from a production well at more than 200°C, producing more than 15 megawatts of thermal output. As of August 2007, the company had just commenced drilling Australia's first commercial geothermal production well and, by the end of the year, hoped to prove productivity large enough to justify building a geothermal power station. In the longer term, its aims are to provide large amounts of base-load renewable energy to the national grid. The known resources from its exploration tenements are large enough to ultimately power much of Australia.

The Australian Marine Quaternary Program

This is a loose association of researchers who have been coming to the Geology Department (now RSES) since the early 1990s when Patrick De Deckker took on the charter to work on the marine Quaternary record of the Australasian region. This was part of a new initiative for the department. Over the years, seven Postdoctoral Fellows have been associated with the group (M. Ayress, T. T. Barrows, F. Gingele, I. Martinez, S. Nees, A. Rathbun and P. Wells), as well as two long-term Visiting Fellows (C. Hiramatsu from Japex, Japan, and the late J.-J. Pichon, from the CNRS in France) and four short-term Visiting Fellows (Professor A. Altenbach from Germany, the late Professor B. Funnell from Britain, Professor H. Okada from Japan and Professor Wang Pinxian from China). These scientists helped advise students and researchers during their stay, and their visits helped shape the directions of research within the group. Seven PhD students (one based elsewhere on campus), one MSc student and six honours students were supervised as part of the program. There are still one honours student and two PhD students in the process of completing their projects.

Major Research Projects

Plate 6.5 Lunch at Darling Harbour sponsored by CSIRO Marine Laboratories following the visit of the RV Southern Surveyor as part of the Geol 3018 Marine Geology Course held in August 2004 in Sydney. From left to right around the table: James Hunt, Chris Wong, Sarah O'Callaghan (staff), Peter Collett (obscured), Shyan Watson, Merinda Nash, Don MacKenzie (CSIRO) at head of table, Helen Byrne (obscured), Alexandra Hickey, Ryan Owen, Cynhia Bolton, Kim Johnston, Nigel Craddy (staff), Sarah Lawrie, Daisy Summefield and Angelina Branco. Not in the photo: Jessica Bowen-Thomas, Malcolm Mann, Steven Petkovski and Patrick De Deckker.
Photo: P. De Deckker

The group includes scientists from China, Colombia, France, Germany, Italy, Japan, New Zealand, Russia, the United Kingdom and the United States, as well as Australia. Many members of the group also participated in four marine cruises led by Patrick on the *RV Franklin*, plus two on the *RV Southern Surveyor* and four on the *RV Marion Dufresne*. In addition, several members of the group participated on other national and international cruises—namely, L. Armand (*Marion Dufresne*), T. Corrège (two *Franklin* cruises), D. Franklin (*Aurora Australis*), F. X. Gingele (*Franklin* and *Marion Dufresne cruises*), E. Gretton (*Franklin*), D. Murgese (*Marion Dufresne*), S. Nees (two *Marion Dufresne* cruises), V. Passlow (*Franklin*), A. Rathburn (*Aurora Australis*), J. Rogers (*Marion Dufresne* and *Southern Surveyor*), A. Yu. Sadekov (*Marion Dufresne* and *Sonne* cruises), M. I. Spooner (two *Marion Dufresne* cruises), P. Wells (Rig Seismic), M. Young (*Marion Dufresne*) and D. Wilkins (*Southern Surveyor*) (Plates 6.5, 6.6 and 6.7).

Patrick and his co-workers now have a large collection of deep-sea cores stored in the Earth and Marine Sciences Building, as well as numerous sub-samples of other cores from our region. A map detailing the location of the cores studied by members of the group is shown in Plate 6.8. This map does not show the location of other cores acquired during four *Franklin*, two *Marion Dufresne*, one *Sonne* and the two *Southern Surveyor* cruises, which are still to be worked on. Ten PhD theses, one MSc and eight honours theses have thus far been completed, and 95 papers published in international peer-reviewed journals, plus 27 papers of a taxonomic nature in similar journals.

Plate 6.6 RV Franklin at Macquarie Wharf, Hobart, January 1994. From left to right: Leanne Armand, Stefan Nees, Patrick De Deckker, Michael Ayress, Tim Barrows, Tony Rathburn, Chikara Hiramatsu and Jean-Jacques Pichon.

Major Research Projects

Plate 6.7 Emptying the box corer on the rear deck of RV Franklin

Geology at ANU (1959–2009)

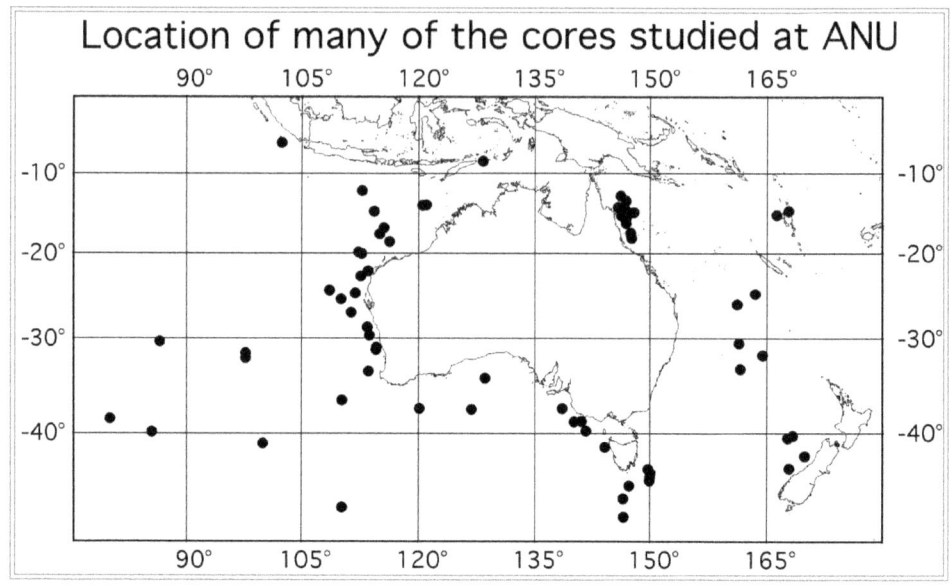

Plate 6.8 Map showing the location of many of the cores studied by members of the group
From P. De Deckker's collection

7
Administrative Work, Honours and Publicity

Faculty Service

Professor Brown was the first Dean of Students after dissolution of the School of General Studies. At the-end-of-year ball, he had to open the dancing with the Student President, a female who was wearing a T-shirt emblazoned 'Make love not war' (in response to the Australian visit of US President Lyndon B. Johnson). Professor Brown also served as external examiner for Newcastle University. Professor Brown, Professor Campbell and Professor Arculus have all been deans of the Science Faculty.

Mike Rickard served as Deputy Dean for several years (1991–95), and Tony Eggleton (1967–68) and Rickard (1979–80) served as sub-deans. In 1981, Mike was a member of the Physics Department Review, and also served as University Marshall for five years (1992–97) (Plate 7.1). George Halford and Mike Rickard were sub-wardens for Burton Hall. Ken Campbell served on the governing body of Burgman College for many years and Tony Eggleton on the board of Toad Hall from 1977 to 1979. Professor Campbell was also a member of the ANU Press Council from 1969 to 1976. Keith Crook was Treasurer of the ANU Staff Association in 1963. Patrick De Deckker and Mike Rickard have served on the Earth Science Library Committee. Patrick served on the Advisory Board of the Kioloa field station, and for several years he was the Graduate School Convener for the Quaternary and Regolith programs. Bear McPhail has served on numerous university committees, including the Geosciences Advisory Group, the Geosciences Education Advisory Group and the ANU Water Initiative Steering Committee. Most staff took a turn on the Science Faculty Board and the Faculty Education Committee and as the Departmental Honours Convener—a position held by David Ellis at present.

Geology at ANU (1959–2009)

Plate 7.1 Dr Mike Rickard as University Marshall

Professional Bodies

Staff played a very active role in support of professional bodies.

Professor David Brown
Professor David Brown was President of the Oceania filial of the International Palaeontological Union in 1969, and was also President of the Geological Society of Australia (GSA) from 1971 to 1973. He was an editor of the GSA Journal for nine years (1964–69 and again in 1976–79). In 1966, David Brown was President of ANZAAS Earth Science Section C in Christchurch, New Zealand.

Professor Ken Campbell
Professor Ken Campbell went to Russia in 1968 for the International Commission on Devonian Stratigraphy; following that, he went to Prague for the International Geological Congress, and was there with Professor Brown when the Russians invaded and aborted the congress. Brown and Campbell later served on the

committee for the twenty-fifth International Geological Congress held in Sydney in 1976. Professor Campbell was also a member of the organising committee for the third Gondwana Symposium. He served as Vice-President and President (1976 and 1977) of the newly formed Association of Australasian Palaeontology and helped set up their journal, *Alcheringa*. Ken Campbell helped organise the Academy of Science's Dorothy Hill Symposium in 1969, and edited *Essays in Honour of Dorothy Hill*. With Gavin Young, he helped organise an international symposium on the Evolution of Early Vertebrates (see Plate 5.1). Ken Campbell raised funds for the Dorothy Hill Award and Lecture at the Academy of Science and he chairs the Award Committee.

Dr Mike Rickard

Dr Mike Rickard was Secretary (1971–73), Public Officer (1977–83) and President (1983–84) of GSA, and, in 1984, sat on the National Committee for Solid Earth Sciences and on the newly formed Geoscience Council. He was also GSA representative to the Commonwealth Institution of Mining and Metallurgy in London in 1969 and 1976, and Executive Secretary of the GSA tectonic map compilation committee (1963–71). In that capacity, he attended a Tectonic Map of the World workshop in Montevideo in 1967. As Australian Convener of WG9 of the International Geodynamics Project (1979–86), he organised the production of several cross-sections of the Lachlan Fold Belt. He was also convener of the Ninth International Basement Tectonics Conference in Canberra (1990). In 1996, he was a member of the organising committee for the GSA's thirteenth convention, held in Canberra.

Professor Bruce Chappell

Professor Bruce Chappell, together with Ken Campbell, visited China on an Academica Sinica Exchange in 1982, and, in 1991, he was convener of the second international Hutton Symposium on granites held in Canberra. He also represented Australia on the International Geological Correlation Program (IGCP) Project 30 'Circum-Pacific Plutonism' for nine years.

Professor Stephen Cox

Professor Stephen Cox was a member of the ARC Fellowship Committee (1997–99); Co-Convener of the Earth Sciences Graduate Program (2000 – present); and Associate Director of Education in RSES from 2008 until the present.

Dr Keith Crook

Dr Keith Crook was Convener of the twelfth International Sedimentological Congress held in Canberra in 1986. He served as a member (including as Chairman, 1986–89) of the Executive of the Consortium for Ocean Geoscience of Australasian Universities (COGS) from 1975 to 1992, and was instrumental in organising Australia's membership of the program. He was a member (1980–82) and Chairman (1982–87) of the IUGS Committee on Sedimentology (1980–82). Keith was co-convener of the sedimentology section for the twenty-seventh IGC, held in Moscow in 1984. In 1998, Keith Crook and John Tipper, on behalf of COGS, were awarded an ARGS grant of $600 000 per annum for three years to support Australasian membership of the Ocean Drilling Program.

Professor Tony Eggleton

Professor Tony Eggleton has been a Council Member of the Association Internationale pour l'Etude des Argiles, and General Chairman of the tenth International Clay Conference in 1993. As President of the Australian Clay Minerals Society for eight years, he was responsible for the development of a constitution and for the society's incorporation.

Professor Patrick De Deckker

Professor Patrick De Deckker has been a member of the following committees: Scientific Committee for the Australian Ocean Drilling Program (1989–92); the Antarctic Research Evaluation Group, Antarctic Division, Natural Environment Subcommittee (1989–92), and Earth Sciences Committee (1995–2000); the Quaternary Science Committee of the Australian Academy of Science (1990–93); Chairman of the Research Group on Ostracoda (1991–94), which is part of the International Palaeontological Association; International Secretary of the Research Group on Ostracoda (1982–88); Executive Committee for the Australasian Quaternary Association (AQUA) (1985–89); the Executive Committee for the Australian Society for Limnology (1986); Secretary of the Sedimentology Specialist Group of the GSA (1988–92); the International PAGES Scientific Steering Committee, associated with the International Geosphere and Biosphere Program (1997–2002); Chairman of AINSE Committee dealing with archaeology and geosciences (since 2004); the Council of the Australian International Ocean Drilling Program (2008).

Patrick's other professional activities, associated with research, include leader of the 'University of the Sea' program (June–July 2005) on board the French vessel *Marion Dufresne*, which provided teaching and research experience to

20 students from the Australasian region. The cruise took place between Papua New Guinea, Indonesia and Australia. He was co-leader of a similar University of the Sea program (February 2006) onboard the *Marion Dufresne*; this time the cruise was held in the Tasman Sea.

Professor David Ellis

Professor David Ellis was leader of UN Educational, Scientific and Cultural Organisation (UNESCO) IGCP Project 236, 'Precambrian events in Gondwana fragments', for several years; during that time he organised meetings in Sri Lanka (1957) and Kenya (1959). In 1996, he was a member of the organising committee for the GSA's thirteenth convention held in Canberra. David was joint convener (1991) for Symposium 31, 'Deep crustal structure of orogenic belts and continents', for the twenty-ninth IGC meeting in Kyoto, Japan. He helped organise (in 1993) a Canberra meeting of the International Association of Volcanology and Chemistry of the Earth's Interior.

Professor Richard Arculus

Professor Richard Arculus was Chief Scientist on the Australian National Marine Facility (*RV Southern Surveyor*) Tonga Eastern Lau Vents Expedition cruise in 2002–03. He also won grants for two cruises in June and November 2004.

Dr Michael Ellwood

Dr Michael Ellwood is currently a member of the Geoscience Advisory Group within RSES. He is also joint academic contact, with Professor Bill Foley, for the Marine Science Major degree and contact for the Marine Geoscience Major degree.

Dr Brad Opdyke

Dr Brad Opdyke has been Australian representative to the following organisations: International Scientific Committee on Ocean Research Working Group 104, 'Coral reef responses to global change: the role of adaption'; South-East Asian Land Ocean Interactions in the Coastal Zone Working Group; International Marine Past Global Changes Study Program Scientific Committee; International Geological Correlation Program (IGCP 404) 'Terrestrial carbon in the past 125Ka' (1996–99); and in 2001–02, he was invited to join the International Geosphere Biosphere Program 'futures' Working Group to help plan the next decade of international marine research.

Since 2002, he has been President of the Australian Sedimentologists Specialist Group of the GSA.

Dr John Walshe

Dr John Walshe served as a committee member of the GSA Specialist Group in Economic Geology (1990–92) and joint convener for the conference on Tectonics and Metallogenesis of the Lachlan Fold Belt (1991).

Dr John Tipper

Dr John Tipper served a term on the Faculties Computing Advisory Committee (1986–89) and was a member of the Consultative Committee to the University Computing Review (1985–86). His external administrative work included the secretaryship of the Consortium for Ocean Geosciences of Australian Universities (1986–89); his time included the successful ARC funding application for Australian membership in the Ocean Drilling Program (jointly with Keith Crook). His other professional activites included membership of the ACT Divisional Committee of PESA (1991), membership of the Accreditation Panel for Geology Courses for the ACT Schools Accreditation Agency (1988–92), and membership of the Merit Promotions Board for the BMR (1991).

Dr 'Bear' McPhail

Dr 'Bear' McPhail served on the executive of the CRC-LEME from 2003 to 2008 and was a member and the Convener of the Environmental Committee of the Australian Institute of Nuclear Science and Engineering (AINSE) from 2002 to 2006. Bear was a session organiser at the Australian Earth Sciences Conference (2002) and the Goldschmidt Conference (2006).

Dr David Tilley

Dr David Tilley was a councillor for the Commonwealth Territory Division on the National Executive of the GSA in 1996.

Dr Andrew Christy

Dr Andrew Christy is Australian representative on the Commission on New Minerals and Mineral Classification, International Mineralogical Association.

Dr **Crook**, Dr **Rickard**, Dr **Tipper**, Dr **De Deckker** and Dr **Opdyke** served on the committee of the ACT Division of the GSA, and most staff served on various local conference committees.

Editorial Work

Staff have edited books and served on a number of editorial boards of international journals (details are outlined in Table 7.1.

Table 7.1 Editorial work undertaken by staff members of the Geology Department

Staff member	Editorial work
Brown, Campbell, Crook, Rickard, White and Conybeare	Contributed to *The Encyclopedia of World Regional Geology*, edited by Rhodes W. Fairbridge (1975)
Brown	Editor of the Journal of the *Geological Society of Australia*, 1964–69 and 1976–79
Arculus	Associate Editor of the following journals: *Geochemical Journal* (1981–90); *American Mineralogist* (1988–90); *Journal of Volcanology and Geothermal Research* (1993–97); *Journal of Geophysical Research* (Solid Earth) (1992–94); *Geochemical Journal* (1999–2004); Editor of *Geology* (Geological Society of America) (2000–02); *Journal of Petrology* (1997–2004); *Journal of Geophysical Research* (Solid Earth) (2005–)
Campbell	Served on the ANU Press Committee (1969–76); Editor with M. Day of *International Rates of Evolution Symposium* (1986, Allen & Unwin); Editor of the *Dorothy Hill Symposium for the Academy of Science*; edited the proceedings of the third volume of *Gondwana Geology* (1975)
Cox	Member of the editorial boards of: *Journal of Structural Geology* (1992–); *Geofluids* (2000–); *Geology* (1995–98)
Christy	Associate Editor: *Mineralogical Magazine* (UK) (1998–); *Central European Journal of Geosciences* (2008–)
Crook	Foundation Member of Editorial Board for *Sedimentary Geology* (1967–87); Editor-in-Chief of *Sedimentary Geology* (1987–2006); Member Editorial Board *Geology* (1983–89)
De Deckker	Edited and co-authored seven journal volumes or books; co-author of two editions of the textbook *Quaternary Environment*; co-chief editor of *Palaeogeography, Palaeoecology, Palaeoclimatology* (1991–2004); on the Editorial Board of this journal, plus the journals of *Paleolimnology* (1989–), *Marine Micropaleontology* (1996–), and *Micropaleontology* (1995–98); board member of the *Journal of Salt Lake Research* (1992–98)
Opdyke	Associate Editor, *Geological Society of America Bulletin* (1997–98); board member of *Geology* (2006–08); Editor, *Geology* (2008–)
Ellwood	Editor, *Journal of Marine and Freshwater Research*
Mavrogenes	Associate Editor, *Journal of Economic Geology* (1999–2004)
Rickard	Senior Editor, *Basement Tectonics 9: Australia and other regions* (1990).
Walshe	Member of Editorial Board of *Mineralium Deposita* (1993–2001); Senior Editor, *Special Issues of Economic Geology: 'The metallogeny of the Tasman Fold Belt system of eastern Australia'* (1995); Co-Editor of a special issue of *Ore Geology Reviews: The conjunction of processes resulting in the formation of orebodies* (1996)

Geology at ANU (1959–2009)

Honours

Member of the department have achieved numerous honours and medal awards (Table 7.2).

Table 7.2 Honours conferred on members of the Geology Department

Staff member	Award
David Brown	Moiety of the Lyell Fund of the Geological Society of London, 1952 Honorary Member of the GSA, 1983 W. R. Browne Medal, 1992
Ken Campbell	Clarke Memorial Lecturer for the Royal Society of New South Wales, 1975 Elected to the Australian Academy of Science, 1983 Mawson Medal and delivered the Mawson Lecture, 1986 Honorary Member of the GSA, 1992 W. R. Browne Medal, 2006
Bruce Chappell	Certificate of Merit, Prince Phillip Prize for Australian Design, 1971 Elected Fellow, Mineralogical Society of America, 1983 Clarke Memorial Lecturer for the Royal Society of New South Wales, 1993 Elected Fellow of the Geological Society of America, 1995 Elected Honorary Fellow of the Geological Society of London, 1995 Mawson Medal from the Australian Academy of Science and delivered the Mawson Lecture, 1998 Elected a Fellow of the Australian Academy of Science, 1998 Centenary Medal for Service to Australian Society, 2003 Honorary Member of Geological Society of Australia, 2007
Richard Arculus	Delivered the fiftieth Clarke Memorial Lecture to the Royal Society of New South Wales, and the Quantum Lecture at Questacon, Canberra, 1991
Steven Cox	Delivered the Ernst Cloos Memorial Lecture at Johns Hopkins University, Baltimore, 2003 Distinguished Lecturer, Society of Economic Geologists, 2007
Keith Crook	Clarke Medal of the Royal Society of New South Wales, 1983 Made Honorary Fellow of GSA, 2008
Mike Rickard	Made Honorary Member and Fellow of the GSA, 1995
Tony Eggleton	Hawley Medal of the Canadian Mineralogical Society (jointly with Professor S. Guggenheim), 1987
John Walshe	Fellow of the Society of Economic Geology, 1985
Patrick De Deckker	Verco Medal by the Royal Society of South Australia for excellence in geology and zoology, 1992 The Australian Society for Limnology Medal in recognition of his work on lakes, 2004 Order of Australia Medal (AM) for his contributions to research and teaching in palaeoclimates, salination and international collaboration, 2007 Christoffel Plantin Medal for contributions to science by a Belgian citizen living abroad, 2008
Gavin Young	Humbolt Awards for palaeontological research at the Museum fur Naturkunde, Berlin, 2000–04
Prame Chopra	ABC Science Media Award Fellowship, Sydney, for the Hot-Fractured-Rock project, 2000

Administrative Work, Honours and Publicity

Dr Reginald C. Sprigg was awarded an Honorary DSc by the ANU in 1980 for his famous work on the late Precambrian 'Ediacara' fossils from Arkaroola in the Flinders Ranges and in the Coorong Lagoon, South Australia. The recipients of the Stillwell Medal awarded by the Geological Society of Australia for best paper of the year in the *Australian Journal of Earth Sciences* are listed in Table 7.3.

Table 7.3 Recipients of the Stillwell Medal

Recipient	Year	Title of paper
R. Both* and K. Williams	1968	The Heemskirk granite in western Tasmania
C. E. B. Conybeare	1970	The solubility and mobility of petroleum in the Surat Basin, Queensland
J. L. Walshe with M. Solomon and C. J. Eastoe	1987	Experiments on convection and their relevance to the genesis of massive sulphide deposits
B. W. Chappell, A. J. R. White and R. Hine*	1988	Granite provinces and basement terranes in the Lachlan Fold Belt
D. H. Moore* et al.	2006	Late Neogene strandlines, etc.
N. Exon, P. Hill (Visiting Fellows) et al.	2006	The Kenn Plateau off NE Australia, etc.

* former student of the department

The members of the department who received the A. B. Edwards Medal for best paper on economic geology in the Australian Journal of Earth Sciences (AJES) are listed in Table 7.4.

Table 7.4 Recipients of the A. B. Edwards Medal

Recipient	Year	Title of paper
R. M. Ahmad* and J. L. Walshe	1990	Wall rock alteration at the Emperor Mine in Fiji
M. J. Rickard and K. McQueen	1996	Structural controls on the Cowarra gold deposits near Bredbo, south-eastern New South Wales
P. Sorjonen-Ward*, Y. Zhang and C. Zhao	2002	Numerical modelling of orogenic processes and gold mineralisation in the south-eastern part of the Yilgarn Craton, Western Australia

* former students of the department

D. A. Brown Medallists

Since 1985, the department has honoured some of our graduates with the award of the D. A. Brown Medal for distinguished achievement. The recipients have delivered a public lecture at the ANU for the presentation ceremony.

Geology at ANU (1959–2009)

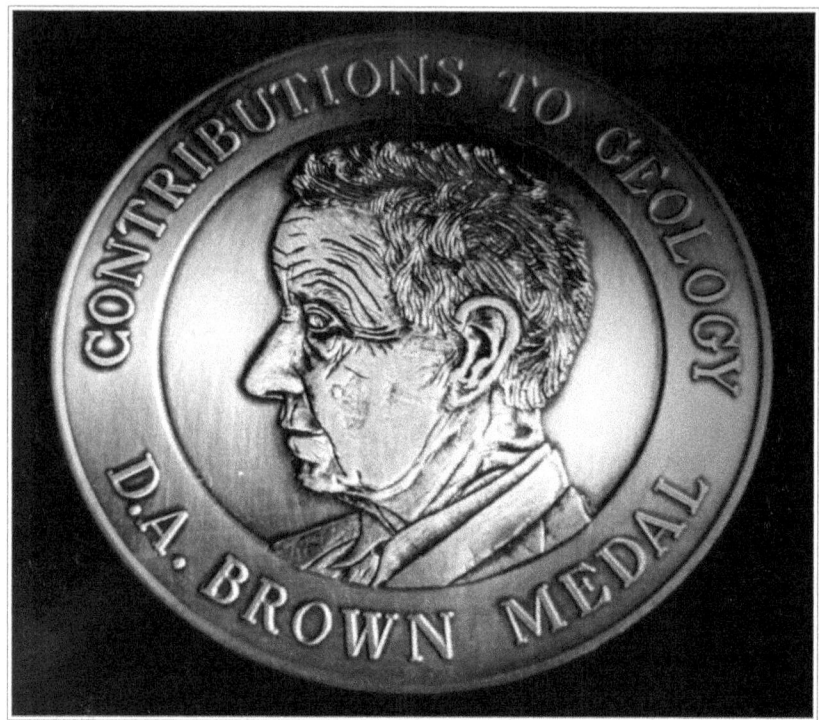

Plate 7.2 The D. A. Brown Medal

Dr Douglas Haynes, 1985

Dr Douglas Haynes graduated from the University of Western Australia in 1966 and completed his PhD at the ANU in 1972. After this, he worked as a specialist mineral-exploration geologist using innovative methods developed during his thesis research. He led a Western Mining exploration team that was responsible for the discovery of the Roxby Downs ore body. His lecture was titled 'Conceptual models in geology and successful mineral exploration'.

Professor Brian D. E. Chatterton, 1987

Professor Brian D. E. Chatterton graduated from Trinity College, Dublin, in 1965 and studied palaeontology for his 1970 PhD at the ANU. Although working mainly on brachiopods at Taemas, New South Wales, he made startling and important discoveries of early trilobite development. He moved to Canada to the University of Alberta (Edmonton) and continued his research in north-western Canada and the Arctic. He became Executive Director of the Canadian Geoscience Council. His lecture dealt with 'Global changes in sea level: their impact on episodic evolution and extinction of trilobites', which was based on his work on the Ordovician–Silurian sequences in the Mackenzie Mountains.

Dr Ian S. Williams, 1989

Dr Ian S. Williams graduated from the Geology Department, ANU, with First-Class Honours in 1974, and studied isotope geochemistry for his PhD at RSES in 1978. After three years at the California Institute of Technology, he returned to RSES to join the ion-microprobe group, which he now leads. He has carried out numerous projects involving the dating of zircons. He lectured on 'Time is of the essence: the changing role of geochronology in deciphering Earth's history'.

Dr Peter A. Jell, 1991 (1992)

After taking First-Class Honours and rugby and rowing blues from the University of Queensland, Peter worked as a curator in the Queensland Museum before coming to ANU to study for his PhD. In 1973, he was given an ANU award to study in the United States. Since graduating, his research has concentrated on Cambrian faunas throughout the world. He has worked as a curator at the Museum of Victoria and as Deputy Director of the Queensland Museum. He has also been President and Editor for the Association of Australasian Palaeontologists. His medal-award talk was on 'New Australian starfish fossils and their friends: what they tell us about evolution and ancient geographies'. This presentation coincided with the official opening of the New Geology Building.

Maurice W. Bell, 1993

Maurice W. Bell graduated from the ANU with honours in 1972 and studied palaeontology for a while. He switched fields to hydrogeology, taking an MSc (*cum laudae*) at Imperial College London and became a successful international consultant on water-resource technology. His lecture on 'Walking on water' recounted incidents and problems encountered with consulting activity in hydrology.

Dr Neil Williams, 1995 (1996)

Dr Neil Williams graduated from the ANU with First-Class Honours in 1969 and received his PhD from Yale University in 1976. He carried out research at RSES from 1976 to 1981, and, in 1980, taught economic geology for the Geology Department. He received the Lindgren Award for his research on the McArthur River (Northern Territory) Pb-Zn deposit. He has held various positions in the mineral-exploration industry, including Chief Geologist with MIM Holdings Limited. He was Foundation Executive Director of the Bureau of Resource Sciences and is currently Director of Geoscience Australia. His medal lecture was 'Society's dependency on geoscience'.

Professor Jill Banfield, 1998 (1999)

Professor Jill Banfield graduated from the ANU with First-Class Honours and, after a year working with Western Mining Corporation, returned to the ANU to study mineralogy for an MSc. She then took a PhD at the Johns Hopkins University, USA, in 1986, and taught as an Associate Professor in the Department of Geology and Geophysics, University of Wisconsin, Madison. Jill is currently at the University of California, Berkeley. Her research into the reactions at mineral surfaces and the nature of interactions between micro-organisms and minerals won her a Mineralogical Society of America Award for young scientists. Her public lecture on receipt of the D. A. Brown Medal was 'Microbial controls on climate, soil fertility, and the environment'.

Dr Ian B. Lambert, 2001

Dr Ian B. Lambert was the first ANU Geology honours graduate in 1963, and he completed a PhD in Geochemistry at RSES in 1967. He joined CSIRO and carried out research on ore deposits and Earth evolution in the Baas Becking Geobiological Laboratory for 18 years. Since 1990, he has held several senior science management positions and is currently at Geoscience Australia, where he leads the National Projects and Advice Group. His lecture was on 'A journey towards sustainable development'.

Professor John M. A. Chappell, 2003

Professor John M. A. Chappell graduated BSc MSc (Auckland) and received his PhD from the ANU in 1973, and is a Fellow of the Australian Academy of Science. He has lectured in geography at the ANU and at McGill University in Canada. From 1974, he was Senior Lecturer then Reader in the Geography Department, ANU, before becoming a Professorial Fellow in Geomorphology and later Head of Department at the Research School of Pacific and Asian Studies (RSPAS). He subsequently transferred to RSES. John Chappell's research has been concerned with sea-level and climate changes, and landscape processes. His lecture was entitled 'Leonardo's legacy and landscape as geology: old wine in new bottles'.

Publicity

The department has participated actively in ANU Open Days, in liaising with schoolteachers and in ACT Career Days. Technical staff have assisted with Open Days. They twice converted the garage and a workshop in the Old Building into a mine, and arranged drilling and panning demonstrations outside. As the New Building was off the beaten track, they obtained permission to paint dinosaur footprints from the central Union Building to the Geology Department to guide potential students and visitors. In spite of using water-based paint, as agreed, the prints took two years to fade.

For the Australian Capital Territory's 1975 'Festival of Australia', we had been entrusted to keep a valuable sample of moon rock, loaned by NASA, and stored in the department's safe over the weekend. When the security escort came to collect it on Monday morning, it was missing, but it was quickly found in the petrology laboratory, where Bruce Chappell was demonstrating it to his class. The department also presented a computer display of moving continents prepared by Research Assistant Lee Belbin. Fossil-fish displays have been arranged for the Opal Museum and the National Aquarium. The tourist pamphlet on the geology of Mount Dromedary prepared by Tony Eggleton has already been mentioned in the section on Student Field Work in Chapter 4 (Plate 4.7).

From 1988 to 1996, the department published its own annual reports (Plates 7.3 and 7.4), which served to attract new students and to publicise our achievements to the exploration companies that employed our students. Patrick De Deckker also produced a mini-disc advertising the department's work, which was distributed to potential students in 2004.

Over the years, the department has achieved major recognition for its accomplishments, with excellent achievements in research grants gained and citation indices for publications. It has developed important relations with employers in the exploration and mining industries. For example, our work in geochemistry and structural geology has had a major influence in providing a basis for further mineral exploration. Together with CSIRO, our studies of the regolith have underpinned the search for buried mineral deposits. Palaeontology, especially work on ancient fishes, has contributed to studies on evolution and has had a major impact on the interest of the general public. More recently, our participation in the growing exploration of the ocean floor by research vessels has contributed to greater understanding of ocean currents and climate change.

Geology at ANU (1959–2009)

The department has always been community oriented. Staff have regularly consulted with schoolteachers, encouraging students to take up teaching, and have run field excursions for teachers and adult education courses.

Plate 7.3 Cover of the 1991 Annual Report. Top right: Leah Moore, Anthony Budd and Mehrooz Aspandiar, leaders of the student excursion to New Zealand. Top left: Waiotapu geothermal field. Bottom left: Eglington Valley. Botom right: Haast gneisses at Fox Glacier.
Photos: Geoff Deacon

Administrative Work, Honours and Publicity

Plate 7.4 Cover of 1992 Annual Report. Top left: Professor Brown plants a ginkgo tree in the courtyard of the New Building, and installs a commemorative plaque on the rock wall of the Old Building (top right). Bottom left: Mike Rickard conducts the opening of the new Geology Department in the Old Botany Building, and Alison McArd from Prospectors Supplies presents the first-year hammer prize to Tim Barrows (bottom right).

Photos: Tim Munson

8
Student Activities

Student Awards

The third-year pass degree is not constrained, so that the amount of geology included varies. This makes it difficult to determine for statistical surveys the actual number of geologists who have graduated from the department. A professional qualification in geology has always been considered to be at the honours level, although several students with pass degrees or graduate diplomas have been successful. Initially, female graduates had trouble finding jobs with mining companies. In 1969, Anne Felton was the first female honours graduate, but, by the mid-1970s, there was a reasonable balance, and 'girls' won several of the major prizes and had little difficulty finding jobs. Photographs from two honours classes are presented in Plates 8.1 and 8.2. Responses to a survey of alumni graduates are included as Chapter 11.

Plate 8.1 Honours year 1987 outside the Old Geology Building. Top: Susie Urbaniak. Middle: Robin de Vries, Kathryn Jagodzinski and Pauline English. Bottom: Michael Conan-Davies, Martin Grant, Matt Yacopetti, Michael Vickery and Phil Jankowski.
Photo: from M. Conan-Davies

Geology at ANU (1959–2009)

Plate 8.2 1993 Geology honours graduates with the ANU Marshall. Left to right top: Claudia Camarotto, Geoff Deacon, Mike Rickard, Allison Britt, Heather Spandler. Bottom: Clinton Rivers, Kathie Barr, Ron Hackney.
Photo: K. Hackney

Table 8.1 Prizes awarded for outstanding achievement by undergraduate Geology students

Year	Prize
First year	GSA Prospectors Supplies (Hammer Prize, see Plate 7.4)
Second year	W. B. Clarke Prize, funded by Geology Department staff
Third year	Anthony Seelaf Memorial Prize for fieldwork Irene Crespin Prize in Palaeontology Geophysics Prize, instituted by Ted Lilley and the Australian Society of Exploration Geophysicists Ampol Exploration Prize for a student entering honours
Honours year	Anton Hales Scholarship to an outstanding honours student, each year, awarded by RSES Western Mining Prize CRA Prize for honours fieldwork projects

The prizes available to Geology students in their undergraduate years are outlined in Table 8.1. Prize winners are listed in Tables 8.2–8.5 at the end of this chapter. University Medals won by students for exceptional merit on graduation are also included. Honours thesis titles are listed in date order in Appendix 2.

Social Activities

There have been a number of family groups enrolled in geology courses: the Creasers (Phil, Jane and Rob), the Spandlers (Carl, Heather, Margaret and Katie), Steve and Leanne Hill, the Stone twins (Richard and Andrew), Chris and Nicky Arndt, and Penny and Alex King. Some children of former students have also studied with us. It is remarkable, if a little surprising, to be asked by a first-year student 'Do you remember my mother?'.

Several students and staff have married or become partners: Louise Donnelly and Theo Walraven; Robin Willoughby and Mel Laurie; Robin Johnston and Dave Fredericks; Ann Kennedy and Bruce Goleby; Elizabeth Rees and Bruce Webber; Sally Rigden and Robert Hill; Jenny Totterdell and Rob Langford; Jon Olley and Vanessa Grey; Pam Fox and Greg Harper; Tim Griffin and Steph Day; Pauline English and Dave Champion*; Sue Coote and Ray Slater; Allison Britt and Ron Hackney; Meg James and Carl Spandler; Laura Dimmer and Steve Boda; Tony Kemp and Karen Alarcon; Helen Degeling and Andrew Tomkins; Chris De-Vitry and Claudia Camarotto; Angela Davis and Fop Vander Hor; Heather Catchpole and Cody Horgan; Anne Felton and Keith Crook. (*Pauline English was allowed to use a departmental vehicle to collect samples from her field area in the Victorian Alps, but only if she took a companion for safety. She chose Dave Champion; the rest is history!*)

The Student Geological Society and social activity have always featured strongly in the life of the department. Students regularly participated in the departmental weekly seminars. A seminar given by Judith Caton in 2003 was unusual as it was on human digestion, but as she felt that this topic was not very 'geological', she arranged to talk for 20 minutes and follow with a tasting of bush food delicacies (prepared by Judith and aided by members of the department). The response, especially from students, was overwhelming, with the D. A. Brown Lecture Theatre packed to standing room only. Free food was irresistible!

Departmental seminars were run as 'Geobabbles' for some years by Chris Jenkins. Chris, Bill Kiene and Ian Robertson made video recordings of one series. For many years, students produced their own newsletter, *Lithenea* (Plate 8.3), which allowed communication of serious and social news, and promoted some comic banter.

Plate 8.3 *Student Geological Society newsletter, September 1984*

Examples are the following 'advertisement' (Plate 8.4), and the observation from *Lithenea* that 'the most significant feature of third-year units is the use of the prefix "C"—this stands for chaos, calamity, confusion, catastrophe, Crook, Chappell, Cameron, Campbell etc'.

Student Activities

Plate 8.4 Wanted dead or alive: Tipper Dr J.C. 'This man is wanted for the wilful destruction of the minds of students who attempted CO8.* Distinguishing traits are an insatiable appetite for Mars bars and a horrendous foreign accent.'
(*CO8 = third-year seminar option)

The staff also contributed to the fun—for example: these three contributions from Tony Eggleton, the department's resident poet and wit.

Geology at ANU (1959–2009)

1. **Mineralogy**

    You are to be imprisoned within a crystal lattice. Write an essay, no greater than one page, on which mineral you would choose and why.

    One of the most gripping aspects of modern mineralogy has been the development of mineral mancipators, popularly known as 'Jail-House Rocks'. The use of crystal prisns began in the days of the Spanish Inquisition, when heretics were put behind cinnabars and tortured with screw dislocation until they confessed. Such places became known as centres of inversion. Unfortunately, the difficulty of pin-pointing the guilty until they would fit into a primitive cell was one facet restricting the use of such prisns. Recent discoveries (Voorhoeve, 1984; Raymond and Flindell, 1984; Smithies, 1982, 1983, 1984) [third-year students—Ed.] have shown that lecturers can be reduced beyond incoherence simply by reading examination answers, and this has led to a resurgence of the crystal incarceration technique.

    The principle of mineral mancipators is one of symmetry; the prisn fits the crime. Initially they were used as lunatic asylums, principally sillimanite and idocrase. Later developments led to thieves being jailed in crookesite or larsenite after being held overnight in polliceite (which contains caesium). Students caught defacing library books go into harmatome, dictators to elbaite, short-term prisoners to minium, indecent topless bathers to bustamout and those on life sentence to witherite or coffinite. The appropriate destination for geology students is lollingite, or on the basis of CO1 fiasconite.

    Considering the coercion put on me to write this thesis, the crystal prisn that I choose is forsterite.

2. **The Students' Geological Society Dinner, 1984**

    Well, those of you who missed the students' dinner
    Should know you missed another students' winner
    The food itself was desper-ately Austrian
    Which didn't stop it being quite pedestrian
    But that was overbalanced by the liquor
    Which freely flowed, and made some speeches thicker
    But not Mike Rickard's, story after story
    Attracted laughs, (or heckles potatory)
    Then Crooky won the Rockies top Award

(A golden hammer with its claw end flawed),
As well as an award for reference shedding
As thickly as confetti at a wedding.
To honour George (who always gets it right)
He won a kit for testing dichroite.
To square the score, the apathetic staff
As our awards, will cut your marks in half.
A quartet and guitar then celebrated
To end with Waltz Matilda R/S rated
While those who still had pangs of hunger, fed
On Catherine's brilliant map of Bunga Head.

3. Staff–Student Cricket Match, 1985: Staff acceptance

You challenge us to cricket?
You mob that miss or snick it?
The Lillie-hearted staffers
Will be the final laughers
As all the student crap'll
Be hit for 6 by Chappell
Right after Kathy Tuffin
Has knocked out all your stuffin
That spin and speed of Campbells
Will leave your stumps in shambles,
And watch out for the demon
New Zealander Ross Freeman
You'll stare amazed as Munson
Relentlessly piles runs on.
The students will be jealous
of the skills displayed by Ellis
And all round cricket ripper
The Botham-like John Tipper,
While none will prove as hardy
As wicket keeper Radi
The guile of spinner Crook
Will write the record book
When Rickard's wrong'n bowls
You'll think your bats have holes
In fact you're sure to lose;
We mark exams for youse.

Geology at ANU (1959–2009)

Occasional T-shirts appeared. The first, 'Grogan man' (Plate 8.5), designed by Rob Goldsmith, was reputed to be modelled on a staff member.

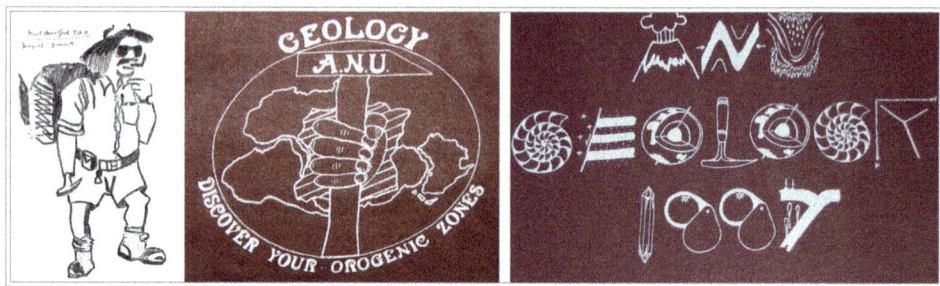

Plate 8.5 Examples of student T-shirt logos
Photos: Mike Rickard

In 1977, Keith Crook was given a Victorian police pith helmet with a metal ANU crest (Plate 8.6); he wore it as a student field-trip hat, and still does in spite of caustic comments in local pubs! On one occasion, his spiel to a student group on Black Mountain was interrupted by two blokes in a passing car yelling 'watch out for tigers'! He was also presented with a walking stick complete with beer holder and bell, which he used to attract student attention until someone filled the bell with peanut butter!

Plate 8.6 Keith Crook in his field hat
Photo: Brad Pillans

Several Geology students played in the 1960s ANU Aussie Rules football teams managed by Bruce Chappell (Plate 8.7). Staff versus student cricket matches, organised by Tony Eggleton, often marked the end-of-year festivities (see Staff acceptance poem, above). Teams were rounded up from throughout the department and sometimes numbered up to 15 a side. One year, we played baseball. Lunch-time footy matches were also common, until a staff member (Dave Ellis) broke an ankle. Some staff and students played tennis, squash and badminton regularly until old age caught up—mainly with the staff members. For many years, there was an annual golf challenge match against Zoology. On one occasion, Euan Reid (a PhD student) led a cross-country skiing expedition to the Snowy Mountains, but a violent storm in the night collapsed our tent and we piled into a small hut, sleeping fitfully on the floor.

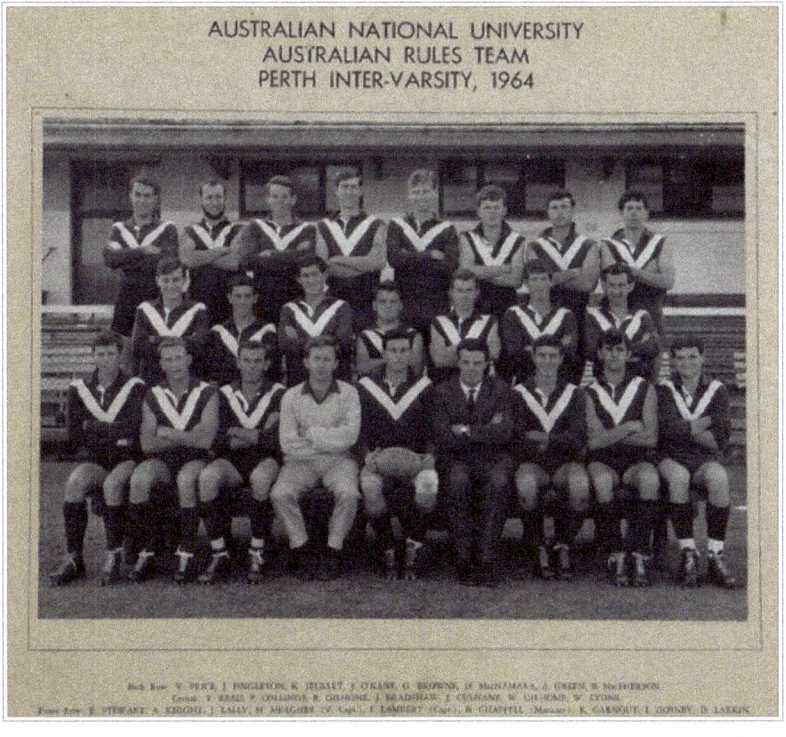

Plate 8.7 ANU Australian Rules team, Perth Intervarsity, 1964. Geology students— middle row, second from left: Peter Collings (now working for Gina Reinhardt's exploration team); second from right: Warren Gilhome (Manager, Shipping, for RioTinto, now retired). Front row, centre: Ian Lambert (now with GA), Bruce Chappell (in suit), team manager, and Ross Garnaut (now Professor of Economics, ANU).
Photo: from Ian Lambert

Staff and some graduate students were housed in flats in Manuka on arrival in Canberra. The Eggletons, Gradys and Links were there at the same time and Albie (Link) provided wood from his field area for the cold-winter-evening card games. Other students were housed in flats at David Street, O'Connor, and more modern ones in Curtin. Some students did it tough. One was caught sleeping in the roof of the Chemistry Department and then moved into a friend's woodshed. This student later gained First-Class Honours.

Staff met regularly on a Friday evening with families at the Staff Club. Early departmental staff dinners were rather formal affairs until some wives introduced dancing (Plate 8.8). Soon after, staff joined in with the student dinner dances (the dancing was livelier, and there was even some cross-dressing; Plate 8.9).

The students made ingenious awards to deserving, and sometimes undeserving, characters. When Mike Rickard became Head of Department, they presented him with his office door—which had been removed to ensure he adopted an open-door policy. Dave Ellis obtained baby photos of many of the students one year and the attempted recognition produced great hilarity. Professor Kalervo Rankama, a well-known Finnish geologist, came to one dinner in 1976, as a visitor, and livened up the evening with a series of off-colour jokes. On another occasion, we booked a bus to a bush pub dinner at Gundaroo, where a good time was had by all. Lee Belbin occasionally flew a few staff and students down to Moruya for a weekend's surfing. On an early morning, rough fishing trip out of Bermagui, everyone got seasick except John McDonald, who originated from central Canada! One student threw a celebratory party after completing his MSc at a woolshed known as 'Piddler's Paradise' on Mountain Creek. The party was enlivened by a Russian visitor ('Vlad the Lad'), who, while chasing a female student, fell from the steps, not knowing that they turned at right angles. The Student Geological Society was also active in ANU Orientation Weeks. On one occasion, they beat Forestry in a wood-chopping competition. They also 'phased out' an astrology presentation by displaying a giant quartz crystal from our museum!

The large class of 1969 has kept in touch since graduating, due largely to the efforts of Anne Felton (Plate 8.10). They held reunions in 1979 and 1989. At the latter, they presented the department with two granite coffee tables for the tearoom—one of S-type and the other of I-type.

Plate 8.8 David Brown and Allan White in party mood
Photo: Ken Williams

Plate 8.9 Cross-dressing: Warwick Crowe and Greg Miles
Photo: from Greg Miles

Some practical jokes are worth mentioning. A flamingo (plastic model) appeared one morning in the flume before a class demonstration. On a trip up north, one student (Euan Reid) expertly climbed the front of the local Bank of New South Wales while awaiting takeaway meals. Reid later climbed a coconut tree in the hotel grounds on Magnetic Island, and was followed by a few others, some of whom skinned their chests when sliding down! On some early camping trips, the 'girls' found their tent door lacings had been changed overnight. Later, most 'girls' rejected separate accommodation and mucked in with the boys! It was Keith Crook's habit to bring each day a hard-boiled egg in his lunch. He routinely dropped the egg on to the floor to crack the shell. On one occasion his boiled egg was replaced with a raw one, creating awed anticipation and then much hilarity as he went through his usual routine.

Geology at ANU (1959–2009)

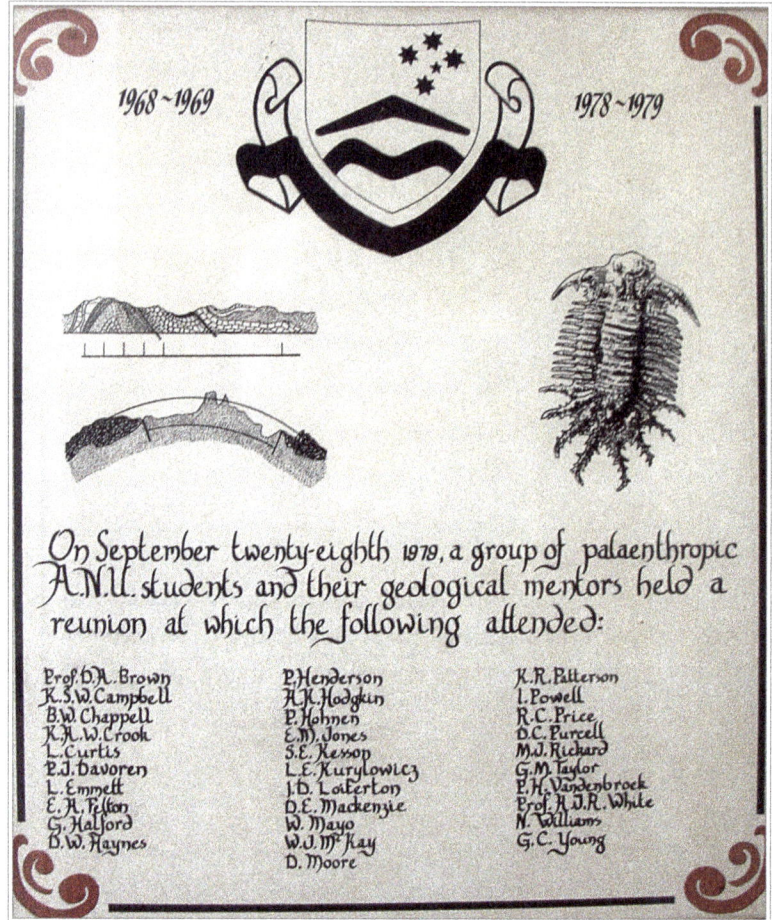

Plate 8.10 Honours Class of 1969 reunion plaque in the Geology Department foyer
Photo: R. Barwick

The department did not escape some student unrest; in 1974, a few Geology students protested against the faculty rule demanding that shoes be worn in science buildings. A crowd of mainly non-Geology, barefoot students occupied the building and interrupted a first-year lecture. The demonstration ended quickly after someone spread drawing pins across the floor of the foyer.

Field Excursions

The highlight of the end of every year was a major field excursion organised by students, with staff helping as necessary. Sometimes they were able to get financial assistance from exploration companies. For one or two years, the students raised money themselves by preparing boxed sets of samples and thin sections of the newly famous S-type and I-type granites. These were an immense success, but sales dropped when they offered a supplementary set of A-types. Perhaps purchasers thought they would be hit with the whole alphabet! They produced and sold clocks made of polished granite gneiss and rock-sample sets for schools. They also printed and sold T-shirts for the GSA's thirteenth symposium, which was held in Canberra.

Plate 8.11 Mt Tarawera, New Zealand: end-of-year field excursion, c. 1992. Top: Anthony Budd. Middle, left to right: Greg Miles, Geoff Deacon, Matt Choquinot, Leah Moore, Allison Britt, Trent Liang. Bottom: Louise Broadbent, Mehrooz Aspandiar, Ron Hackney.
Photo: Greg Miles

These field excursions visited the following places (with leaders' names following): New England, twice (Chappell); Tasmania, twice (Eggleton and Munson); Broken Hill (McDonald); Alice Springs (Rickard and Halford); Western Australia (Allison Britt); Flinders Ranges (White and Munson); Fiji (Rickard, Crook and Walshe); Hawai'i (Anne Felton); New Zealand, twice (Feary, Cameron, Moore and Anthony Budd) (Plate 8.11); New Caledonia (De Deckker); the Pilbara (McDonald); Chile (Arculus) (Back piece). On one New England excursion, we were accompanied by several visiting Chinese geologists, who were bemused when we all went to the local drive-in and watched Star Wars lying outside on our camp beds!

Geology at ANU (1959–2009)

Graduate Destinations

Over the 50 years, more than 300 honours and 40 Graduate Diploma students, 95 PhD and 46 MSc students have graduated from the Geology Department at the ANU. During boom employment times, several companies regularly visited the department seeking the best students. Many students have found employment with the BMR (AGSO, GA) or with CSIRO, State geological surveys or museums. Most have gone to mining companies working largely in exploration; some have risen to high management positions. A group of three graduates (Richard Hine, Fiona Morrison and Donal Windrim) found themselves all working on a special research topic for BHP-Utah producing a tectonic map of the Tasman orogenic system under the leadership of consultant tectonicist, Peter Coney, from the University of Arizona. A few students with good maths backgrounds found employment in geophysics with oil companies, and one or two in hydrogeology.

Many of our students have taken up academic positions: M. Stauffer, B. Chatterton, R. Creaser (University of Alberta, Edmonton); P. Johnson (Calgary); M. Rhodes (University of Massachusetts, Amherst); N. Arndt (Universities of Saskatoon and Grenoble, France); Jill Banfield (University of California, Berkeley); I. Smith (Auckland University); G. Davidson (University of Tasmania); J. Cooper, D. McKirdy, V. Gostin, R. Both, J. Foden and A. Schmidt Mumm (University of Adelaide); D. Legg (Charles Sturt University, Wagga Wagga); A. Link (RMIT, Melbourne); W. Collins (University of Newcastle and James Cook University, Townsville); C. Marshall (Museum of Comparative Zoology, Harvard); D. Christie (University of Alaska); T. Griffin (La Trobe University); Penny King (Universities of Western Ontario and New Mexico); S. Fermio (SAIT); B. Jones (University of Wollongong); R. Price (Universities of La Trobe and Waikato, New Zealand); B. Pillans and J. Chappell (ANU); G. Taylor and Leah Moore (University of Canberra); R. Hackney (Kiel University, Germany); M. Aspandiar (University of Western Australia, Curtin); Janet Hergt (Melbourne University); I. Martinez (EAFIT, Medellin, Colombia); G. Deacon (Murdoch University); T. Correge (University of Bordeaux); A. Rathburn (Indiana State University and Scripps Institute of Oceanography); A. Kemp (University of Bristol and James Cook University, Townsville); C. Spandler (James Cook University, Townsville); A. Hack (ETH, Zurich); I. Ferguson (University of Manitoba, Winnepeg); J. Olley (Griffith University); A. Tomkins (Monash University); J. Peters (West Australian School of Mines).

Student Activities

ANU graduates are working, or have worked, overseas in Antarctica, Argentina, Bangladesh, Belgium, Borneo, Canada, Chile, Colombia, Congo, Czech Republic, Estonia, Finland, Fiji, France, Ghana, Germany, Greece, Greenland, Guyana, Holland, India, Indonesia, Iran, Italy, Korea, Liberia, Mali, New Guinea, New Zealand, Namibia, the Netherlands, Nigeria, Norway, Mauritania, Oman, Peru, the Philippines, Russia, Saudi Arabia, Senegal, Sweden, Thailand, Timor, the United Kingdom, the United States and Yemen. While most graduates have found employment in the mineral or petroleum industries, or lately in environmental fields, geology provides a good general science background and some graduates have found employment in diverse fields such as: schoolteaching, IT companies, environmental agencies, resource economics, public relations, law of the sea, Australian Security Intelligence Organisation (ASIO) (remote sensing), Australian Nuclear Science and Technology Organisation (ANSTO); as well as in more exotic fields such as: accounting, antiques, journalism, publishing, boat building, financial advice, fisheries, real estate, shipping, the mint, art, religion, opera, child care, politics, wine making, farming, violin making, and even a union representative for sex workers!

Writing references for students has become an artform, especially when one realises that some miserable achiever will shortly earn more than an academic. Mike Rickard was once interviewed by ASIO (Remote Sensing Section) on behalf of a student applicant. Among other strange and personal questions asked was why the 'girl' had been the only student to attend a lecture by a communist Chinese professor at a recent ANU conference. It turned out she was a student helper working the projector, and found the lecture boring. But how did they know?

Geology at ANU (1959–2009)

Table 8.2 Geology student prizes

A. First-year geology. Three prizes are awarded to first-year students: the Geological Society of Australia Prize, the Professional Officers Association Prize (POP), which is awarded to science departments in turn, and the award of an Estwing geological hammer and Gfeller holster by Prospectors Supplies (PSA).

1959 H. F. Doutch and POP	1988 D. Bush
1960 J. Cleary	A. Goody
1961 N. Henderson	1989 S. Love and PSA
1962 A. Capp and POP	1990 N. C. Davis
1963 M. J. Carr	T. D. Andrews
1964 J. I. Raine	J. Hackney PSA
1966 R. C. Price	1991 R. Bartel
1968 D. L. Gibson	T. T. Barrows PSA
1969 D. J. Holloway	1992 F. Holgate
1971 G. R. Heys	1993 M. James
S. A. Gibbins and POP	I. Mullen PSA
1972 A. J. Williams	1994 M. Chorley
1973 S. M. Rigden	L. Zimmitat PSA
1974 B. R. Golby and POP	1995 T. Martin and PSA
1975 A. Butterfield	1996 K. Worden
1976 R. M. Johnston	I. McGufficke PSA
1977 I. J. Ferguson	1997 M. Fitzell and M. Fit
1978 J. F. Banfield	1998 R. O'Leary
1979 S. M. Howit	P. Alcorn
C. R. Marshall	1999 L. A. Schapel
1980 A. N. Boston	2000 J. R. Edwards
B. S. Turner	L. E. Richardson PSA
1981 J. G. Downing	2001 F. Probert
1982 O. L. Raymond	B. Maguire PSA
1983 S. Sheppard	2003 H. Mason PSA
1984 M. J. Vicary	2004 J. McDonald
1985 G. R. Hunt	2005 S. Holden
1986 A. G. and R. I. Stone and POP	A. Kay PSA
1987 M. C. Hunt	2006 K. Cummin
	A. Lukomskyi PSA
	2007 J. Bennett
	M. Pittard PSA

Student Activities

B. Second-year geology: the W. B. Clarke Prize donated by the Geology Department staff to the best student in second year

1964 M. J. Carr	1989 A. Goody
1965 D. P. Legg and J. I. Raine	1990 S. Love
1967 G. C. Young	1991 M. Bambrick
1968 A. G. Seelaf and W. H. Oldham	1992 T. T. Barrows
1969 R. Hine and D. L. Gibson	1993 F. Holgate
1970 D. J. Holloway	1994 C. Devlin
1971 J. T. Cameron	1995 M. Chorley and A. Hack
1973 A. J. Williams and L. P. Greer	1996 N. Kositcin
1974 S. M. Rigden	1997 C. Leslie
1975 W. J. Collins	1998 C. Gillespie-Jones
1977 R. M. Johnston	1999 A. A. Kalinowski
1978 F. Morrison	2000 J. Wykes
1979 J. F. Banfield	2001 J. R. Edwards
1980 D. J. C. Varkevisser	2002 G. Myers and C. Southby
1981 R. A. Creaser	2003 S. L. Lawrie
1983 O. L. Raymond	2004 S. Hue
1984 S. Sheppard	2005 S. Biddlecombe
1985 S. M. Urbaniak	2006 L. Soroka
1986 C. R. Newman	2007 K. Cummin
1987 G. Deacon, M. B. Mills and E. J. Stanner	
1988 W. Woodhouse	

C. Third-year prizes

1. Anthony Seelaf Prize for fieldwork, donated by the Seelaf family in memory of their son, who was accidentally killed in New Guinea

1979 W. Prohasky	1994 F. Holgate
1980 G. E. Wheller and D. J. C. Varkevisser	1995 C. Steele and M. James
1981 N. C. Robertson	1996 A. Hack
1983 S. J. Fermio and J. M. Olley	1997 P. Jain and M. Holzapfel
1985 T. A. Bennetto and A. O. Piiroinen	1998 R. Lacey
1986 M. A. Grant and C. M. Yacopetti	1999 C. Gillespie-Jones
1987 D. M. Coad and G. D. Howells	2000 S. Zasiadczyk
1988 G. Deacon	2001 M. Worthy
1989 B. Jones	2002 N. Tailby
1990 A. Goody and P. King	2003 C. Southby
1991 L. Mitchell	2004 T. A. James and H. Byrne
1992 R. Hackney and M. Woodbury	2007 J. Lee and J. Garvey
1993 T. T. Barrows	

Geology at ANU (1959–2009)

2. Irene Crespin Prize (Irene Crespin was a BMR/AGSO palaeontologist for many years). The prize bequest is awarded for palaeontology.

1982 M. J. Jones	1998 P. Payne
1985 K. J. Driver	1999 A. A. Kalinowski
1986 P. Thomas	2000 J. R. Edwards
1987 G. Deacon and E. J. Stanner	2001 J. Bowen-Thomas
1988 G. Hunt	2002 C. Gouramanis
1989 C. Nipperess	2003 S. L. Lawrie
1990 D. Franklin	2005 G. Nash
1992 G. Sargent	2006 L. Soroka
1993 R. Bartel	2007 K. Cummin
1995 K. Warren	
1997 S. Dimitriadis	

3. The Geophysics Prize, donated by the Australian Society of Exploration Geophysicists (ACT Branch). This prize is awarded to the student who achieves the best result in geophysics in that year (geophysics is not taught every year).

1996 G. E. Manning	2000 K. Dalby
1997 D. W. Hackney	2002 P. McLean
1998 C. Farmer	2004 C. Thompson
1999 A. Johnston	2006 A. Bigault

Student Activities

D. The Ampol Prize for a student entering honours year, and the Western Mining (WMC) and Conzinc Riotinto (CRA) awards for honours fieldwork

1961 D. J. Belford	CRA–C. Rumble
1965 B. J. Hocking	WMC–S. Bygrave
1967 J. I. Raine	1990 A. Goody
1968 N. Williams	CRA–W. Crowe
1970 R. Hine	WMC–M. Gordon
APEA Conference Award C. B. Tassell	1991 S. Love
1971 D. J. Holloway	1992 R. Hackney
1977 G. R. Mangold	CRA–P. King
1979 F. Morrison	WMC–A. Shepherd
1980 J. F. Banfield	1993 K. Patrick
1981 C. L. A. Killick	1994 Not awarded
1982 A. B. Boston and B. S. Turner	CRA–K. Barr
1983 R. A. Creaser	1995 Not awarded
1984 C. L. McCormack and O. L. Raymond	CRA–T. T. Barrows
1985 S. Sheppard	WMC–F. Holgate
1988 G. Deacon and J. Stanner	1996 Not awarded
CRA–M. Grant	CRA–L. Foster
WMC–P. English	WMC–C. Devlin
1989 M. Hope and W. Woodhouse	1997 Not awarded
	Records not complete

Table 8.3 The K. S. W. Campbell Prize, awarded to students or staff in honour of Professor Ken Campbell for an outstanding contribution to teaching

1993 C. L. Moore
2000 H. McGregor
2002 H. Bostock
2003 F. Holgate
2005 A. Christy
2007 B. Harrold

Geology at ANU (1959–2009)

Table 8.4 University awards

A. ANU Medals awarded for exceptional achievement at honours level

1974	J. M. Kennard
1979	R. M. Johnston
1991	A. Goody
1992	S. B. Love
1994	M. Carey
1995	F. Holgate
1997	A. C. Hack
2003	J. L. Wykes

B. The Priscilla-Fairfield-Bok Prize for excellent achievement in science by a female student

1986	S. M. Urbaniak
1992	C. Camarotto

Table 8.5 Robert Hill Memorial Prize, in honour of a former student and RSES Fellow

1993	A. Stamp	2000	T. T. Barrows*
1994	D. Post	2002	E. Hendy
1995	M. Bourke	2003	N. Abram
1996	S. Hill*	2004	F. Herman
1997	G. Batt	2006	F. Reith
1999	D. Sinclair	2007	K. E. Fitzsimmons and D. J. Robinson

* ANU honours student

9
The New Millennium

In 1998, funding cuts led an effort to increase student loads. From 1995 to 1999, student load increased considerably, while at the same time full-time academic staff numbers were reduced from eight to six. As a result, student–staff ratios declined from 16:5 to 37:1. In addition, several administrative duties were downloaded from central to departmental levels.

From 2000 on, the department underwent many changes. It became part of a wider College of Science, changed its name to the Department of Earth and Marine Sciences (DEMS) in 2004, and then merged with RSES in late 2007. There were four heads of department over this time: Richard Arculus, David Ellis, Patrick De Deckker and Bear McPhail. In 2008, after the merger with RSES, Professor Brian Kennett, the Director of RSES, took over the whole school, and Professor Stephen Cox was appointed as Associate Director Education to supervise the undergraduate program. Another major development affecting the Faculty of Science was the establishment of the Fenner School of Environment and Society as a joint centre for research and environmental studies.

A prospectus for research and research training in Australian Earth Science, *Towards 2005*, was prepared in 1998 by a working party of the Australian Geoscience Council for the ARC. This report recommended, inter alia, an increase in practical training and an emphasis on environmental studies. The department responded by forging closer links with the Australasian Institute of Mining and Metallurgy (AusIMM) through the good offices of Peter Hancock (a Visiting Fellow). The department also reversed its previous stance and allowed AusIMM to accredit ANU courses. AusIMM urged the university to maintain the independence of the Geology Department, and, with the Minerals Council, has been steadfast in urging the government to increase its funding for minerals sector-related courses to the same per-student level as it provides for agricultural courses.

In 2000, Peter Hancock raised funds from South African and Australian companies to establish the AusIMM Australia–South Africa Minerals Scholarship for ANU students to visit South Africa to carry out honours projects, and to gain work experience in a developing country, to travel and meet with industry professionals and academics to advance their own personal and professional development. The aims of this scholarship were to identify, encourage and nurture the attributes of professional geologists of the future who could rise to positions of leadership in the industry. In 2008, Dominique Tanner was the first ANU student to win the prestigious Sir Frank Espie AusIMM Scholarship—one of only three awarded nationally each year from hundreds across the fields of geoscience, engineering, mineral processing and environmental science.

The ANU students formed their own junior branch of the AusIMM, the 'Brindabella Student Chapter', in 2000, under student President Chris Gunton. AusIMM has also run regular careers nights and geo-trivia nights for the students and it supports economic geology field excursions and participation in the annual Central West Mining Forum in Orange. The students also initiated a 'ready for work' course in 2008 with intensive instruction in mining and exploration techniques. This was supported by several companies and AusIMM, and might be extended in future. AusIMM also ran a joint ANU–University of Wollongong mineral venture in which 35 highschool students were taken on a 10-day tour of mines and mineral sites around the Hunter region and western New South Wales in an endeavour to attract more students.

Course Structure

Staff from DEMS and RSES applied for and won a major innovations grant. They were later joined in this initiative by members of BOZO. This led to a change in emphasis, with the inclusion of marine science and a change in the department's name to Earth and Marine Sciences to reflect the broader scope of the courses offered. A new degree—a Bachelor of Global and Ocean Sciences—was introduced with a high UAI-entry requirement (Plate 9.1).

The whole undergraduate curriculum in Earth and Marine Sciences was reviewed in 2006, and this was followed by extensive changes. The degree structure changed to an eight/eight/eight pattern, with four units each semester, instead of three, and the new units were reduced in size to accommodate three lectures and one three-hour practical session a week. Honours pathways involved additional and more advanced work. The department's new curriculum (Table 9.1) begins with two first-year courses in collaboration with the Fenner School of Environment and Society (a combination of the School of Resources, Environment and Society and the Centre for Resource and Environmental Studies). There are six second-year geoscience-related courses, plus a second-year course in environmental chemistry. At third-year level, there are eight semester courses, plus three field courses (field geology, coastal environmental earth sciences, and carbonate reefs). New courses include geophysics, groundwater, global cycles, marine bio-geochemistry, and ocean and atmosphere modelling. Students choose majors in Geology, Double Geology, Environmental Geoscience, Marine Geoscience, and Marine Palaeontology—the first three of which are accredited by the AusIMM. These involve lecturers from outside the Geology Department, especially from RSES and BOZO. The department also contributes to the Minerals Tertiary Education Council (MTEC) program for honours and masters students in participating universities. A specific field geology course (the old C01) has been dropped, but field mapping has been included in the first-year coastal field trip and Professor Steven Cox has been involved in teaching field geology with Brad Opdyke at second and third-year levels.

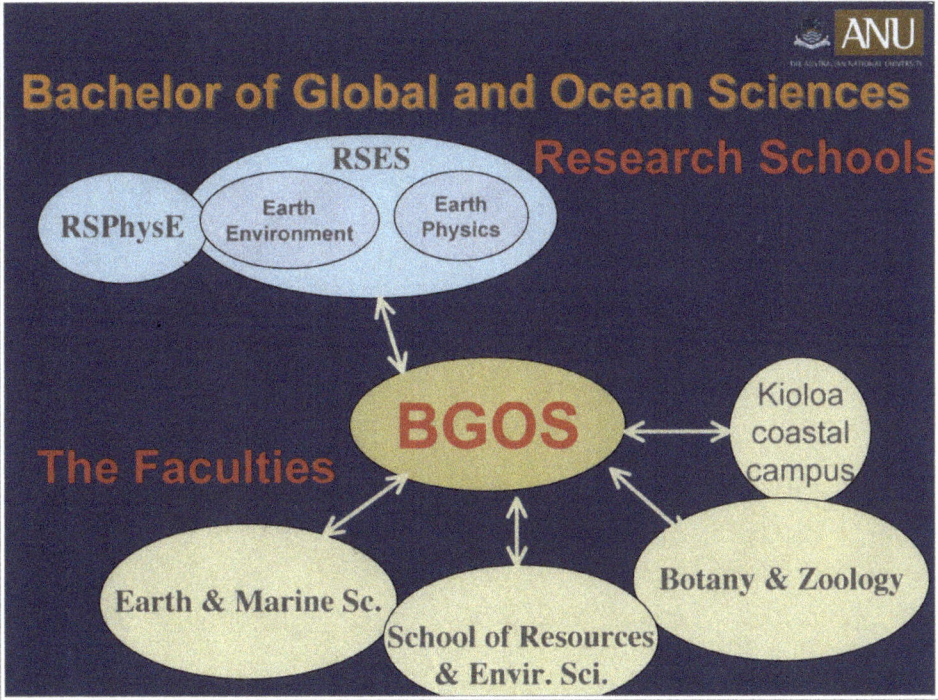

Plate 9.1 Bachelor of Global and Ocean Sciences

Table 9.1 Geology courses offered in 2006–07

Semester 1	Semester 2
Geol 2012 Introduction to Structure and Field Geology (Cox and Opdyke)	Geol 1006 Blue Planet: Introduction to Earth System Sciences (Mackey and Opdyke)
	Geol 2015 Chemistry of Earth and Oceans (Arculus and Rubatto)
Geol 2014 Surficial Processes, Source to Sink (Opdyke)	Geol 2019 Marine Palaeontology and Evolution of Life on Earth (De Deckker)
Geol 2017 Mineralogy (Ellis)	Geol 2020 The Lithosphere (Arculus and Ellis)
Geol 2018 Geophysics# (RSES staff)	
Geol 3002 Structural Geology and Tectonics# (Cox)	Geol 3007 Economic Geology (Mavrogenes)
Geol 3019 Carbonate Reef Studies (De Deckker and Opdyke)	Geol 3025 Groundwater (McPhail)
Geol 3022 Planetary Geology (Arculus and Ireland)	Geol 3026 Environmental and Regolith Geoscience (De Deckker and Pillans)
Geol 3023 Marine Bio-Geochemistry (Eggins and Ellwood)	Geol 3027 Global Cycles and Palaeoecology (Opdyke)
Geol 3024 Magmatism and Metamorphism (Herman)	Geol 3028 Coastal Environmental Earth Science (Ellis and Beavis)
Geol 3030, 3031 and 3050 Advanced Seminar units	Geol 3030, 3031 and 3050 Advanced Seminar units

offered in alternate years

The numbers of undergraduate students have declined to a worryingly low level over this period, although the number of students in honours classes has been high (16 in 2001; Plate 9.2), and the numbers of graduate students have increased. Overseas links have enabled some students to study in France, China and South Africa. Overall, the course offerings have been increased. A new course in planetary geology was given by Professor Arculus and Professor Taylor. A new field course has recently been offered at Broken Hill by Professor Gordon Lister from RSES.

Current Research

Current research directions are closely aligned to the 'Australian Academy of Science National Strategic Plan for the Earth Sciences' (2003). The department's research effort increased and cross-cooperation with RSES—which had always been good—expanded markedly, especially as two staff (Cox and Mavrogenes) were joint appointments.

The New Millennium

Plate 9.2 *The Geology Department's largest honours class—in 2001*

The CRC-LEME grew to be an important section of the department and its operation was extended for a further seven years, with a grant of six new PhD students. Dr Bear McPhail, who was appointed Group Leader after Professor Eggleton's retirement, established a new regolith program involving low-temperature aqueous geochemistry, groundwater hydrology and geo-microbiology. Two new Research Fellows were also appointed: Dr Dirk Kirste, a hydrogeochemist, and Dr Sue Welch, a geo-microbiologist (see Chapter 6, CRC-LEME).

A new Department of Education, Training and Youth Affairs (DETYA) science lectureship, funded by the Minerals Tertiary Education Council, appointed Dr Ian Roach to teach regolith courses. Maree Coldrick (the DEMS Administrator) was seconded part-time as the Secretary of LEME in an attempt to reduce costs. In spite of its success over many years—and its importance to Australian science—it was terminated in 2008.

Success in obtaining research grants boomed, and marine research with international cruises increased with Professor De Deckker and Professor Arculus and Dr Opdyke taking part. The department was successful in its bid to host the office succeeding the Australian Ocean Drilling Program for three years from 2004. These cruises produced newsworthy reports of new discoveries of undersea volcanoes (see Chapter 6).

Two large ARC Discovery Grants were awarded in 2002—one for Professor De Deckker (shared with Professor J. Dodson, University of Western Australia), and the other for Professor Arculus and Dr Mavrogenes (shared with RSES). Professor De Deckker is the leader of a group involving the ANU, several other universities and government departments in an 'Ocean Discovery Network'. An ARC Discovery Grant to Professor Ellis and Research Fellows Dr Ulli Troitzsch and Dr Andrew Christy led to a patent for a new semiconductor (zirconium silicate). Dr Prame Chopra was awarded an ABC Science Media Fellowship and spent time in Sydney with the ABC.

Other research partnerships were established in the Cooperative Research Centre (CRC) for Antarctic Climate and Environment (ACE), the Centre of Excellence for Ore Deposits and Mineral Exploration in Tasmania (CODES), partnership in the Centre of Excellence for Coral Reef Research, strong linkages with GA involving advisory roles and shared facilities, cooperation with CSIRO Exploration and Mining for jointly funded positions in ore-genesis studies, and many successful joint bids with the Universities of Canberra, Wollongong, Curtin and Melbourne for major equipment. These research projects are outlined in more detail in the staff research reports (Chapter 5).

Building and Equipment Support

Some recent changes to the building have occurred. The seminar room named for the founding professor was converted to an honours office and the name transferred to the remodelled lecture theatre on the ground floor. In 2008, the New Geology Building housing the undergraduate teaching section of RSES was named the 'David Brown Building' (Plate 3.7). Upstairs, the Geology II and III laboratories were also refurbished. The office space in the Zoology Building housing our palaeontologists—and for a while the hot-rock project—was relinquished in 2008, and the Visiting Fellows relocated to the main building. Dr Wyborn and Dr Chopra resigned to direct the commercial operations in geothermal energy.

The geochemical laboratories on the ground floor were rearranged. The INNA instruments were sold and new spectrometers were housed in their place (ICPAES: inductively coupled plasma atomic-emission spectrometer, and

ICPMS: inductively coupled plasma mass spectrometer). These were set up with the help of two Research Fellows: Dr T. Eszat and Steve Eggins. A new thin-sectioning and rock-grinding machine was purchased, and the department was part of a consortium to gain a new electron microscope. The Anutech support staff were retrenched and the analytical facilities are now managed by Dr Andrew Christy and Linda McMorrow. Val Elder has continued her sterling voluntary efforts to sort out the museum holdings. Support staff were further reduced by allocating Sarah O'Callaghan part-time to the School of Resource and Environmental Management (SREM) (Fenner School), and then to a full-time position. Currently, the department has only nine technical support staff, compared with 15 in the early days!

The Departmental Review of 2007

A major review of the department chaired by Professor G. Govett of NSW University took place in May 2007. Important recommendations were that

1. the department be combined with RSES within the College of Science, but maintain its location and have a separate education budget and leader
2. the research program should incorporate the distinctive geological identity of current DEMS research
3. the postgraduate programs should be integrated with those of RSES
4. laboratory facilities and equipment should be integrated and rationalised to achieve efficiency
5. following the demise of LEME, the department should re-evaluate its future research directions.

The committee commended the staff of DEMS for the talent and commitment they bring to their teaching program and outstanding support for their students—'a model for the faculty'. They expressed concern, however, about the number of courses offered and that the integration of geology and environmental geology at first-year level did not meet student approval. They were also critical of the lack of geophysics in the curriculum and the lack of prerequisites in maths and physics. There was also a concern that outreach to local schools and industry was inadequate and should be enhanced. Finally, the committee recommended that the department develop operations that increased funding and student enrolments, budgeted better for support services and considered staff rearrangements to effect savings.

10
The Reunion and the Future

A reunion dinner marked the first 50 years of the Geology Department at the ANU. This was a celebration of all that had been achieved by members of the department during those years and it marked the beginning of a new era for the present staff and students.

Plate 10.1 Heads of department past and present. From left to right: Professor Brian Kennett, Bear McPhail, Professor Ken Campbell, Professor David Brown, Mike Rickard, Professor Patrick De Deckker and Professor David Ellis (Professor Richard Arculus was overseas).

The reunion dinner was held in the Great Hall at University House on Saturday 7 March to celebrate the fiftieth anniversary of the setting up of the department. Four of the original 1959 class were present as were all but one (who was overseas) of the heads of department (Plates 10.1 and 10.2). One hundred and forty-one alumni and partners attended (Table 10.1). Professor Brian Kennett, as current Director (RSES), welcomed all the participants and explained that the Geology Department had changed its name to Earth and Marine Sciences and recently had been merged with the Research School of Earth Sciences. The Geology Building had been renamed 'the D. A. Brown Building' (No. 47). Mike Rickard read out some apologies and greetings, and, in proposing a toast to the department, emphasised the ongoing, beneficial and inspirational role of the Founding Professor, David Brown. Responses were made from the first honours graduate, Ian Lambert, and the first female honours graduate, Anne Felton, who reminisced on their student days. Professor Brown responded with his thanks and mentioned the enormous contribution to the success of the Department of Geology made over the 50 years by his staff colleagues and many others. A draft copy of this history manuscript was circulated.

Plate 10.2 *Early students and staff at the reunion dinner, University House, ANU. From left to right: John Cooper (student), Alex Grady (Demonstrator), Jan Grady, Judy Day, Keith Crook (lecturer, now Visiting Fellow), Bob Day (Demonstrator), Anne Felton, (student, now Visiting Fellow), George Halford (student and Museum Curator), with wife, Betty (hidden), and Fred Doutch (student).*

In spite of ongoing criticisms and the difficulties of balancing teaching and research activity, Professor Brown and all the staff who have worked in the department over the years can take pride and pleasure in the fact that today the impact of its teaching has produced successful professionals in so many fields (see Alumni News, Chapter 11), and that its research, publication and citation record is second to none. Now, at the end of a half-century, the Geology Department merges with the RSES to form a world-best operation for the future.

This is not an end; it is a beginning—of the next 50 years or more, which will be so different from the first 50. It is fitting that the first 50 years in the evolution of the Geology Department (now the undergraduate teaching arm of RSES) should be marked by a significant event: the reunion dinner. We hope that the century of the department will be marked in a similar way and that the next 50 years will be as productive and as much fun as the first.

Table 10.1 Attendees at the Geology Reunion Dinner (7 March 2009)

Charlotte Allen	ANU	Ron Hackney	GA	David Purcell	Qld
Leanne Armand	Tas.	George Halford	NSW	Christine Purcell*	
Andrew Barrett	Sydney	Betty Halford*		James (Ian) Raine	NZ
Richard Barwick	ANU	Karen Higgins	GA	Gerry Reinson	Canada
Simon Beams	Qld	David Holloway	Vic.	Barbera Reinson*	
Sara Beavis	ANU	Angela Hume	ACT	Mike Rickard	ANU
Harrold Berents	WA	Nigel Johnson	WA	Jan Rickard*	
Julie Berents*		Peter Jones	ANU	Clinton Rivers	WA
Allison Britt	GA	Mavis Jones*		Mali Rivers*	
Margaret Bromley*		Alexksandr Kalinowski	GA	Heather Robinson*	
David Brown	Sydney	John Kennard	GA	Jane Rodgers	Vic.
Pat Brown*		Brian Kennett	ANU	Judith Shelley	ANU
Cathy Brown	GA	Chris Klootwijk	ANU	Michael Shelley	ANU
Wally Bucknell	Sydney	Nerida Knight	ACT	Collette Shelley*	
Ken Campbell	ANU	Janice Knutson	NSW	Raymond Slater	Qld
Daphne Campbell*		Ian Lambert	GA	Sue Slater	Qld
Felicity Chivas	ACT	Elisabeth Lambert*		Ian Smith	NZ
Bob Close	Qld	Peter Lang	ACT	Peter Sorjonen-Ward	Finland
Michael Conan-Davies	Vic.	Anne Lang*		Peter Stuart-Smith	ACT
John Cooke		Andrew Lawrence	WA	Trish Stuart-Smith*	
John Cooper	SA	Richard Lesh	Qld	Graham Taylor	ANU
Stephen Cox	ANU	Ted Lilley	ANU	Helen Taylor*	
Elizabeth Cox*		Penny Lilley*		Gordon Taylor	ACT
Nigel Craddy	ANU	Ian Loiterton	ACT	Ross Taylor	ANU
Phil Creaser	ACT	John Long	Vic.	Graeme Torr	
Keith Crook	ANU	John Magee		Elizabeth Truswell	ANU
Garry Davidson	Tas.	Doug Mason	SA	Simon Veitch	NSW
Robert Day	Qld	Jo Mason	SA	John Walshe	WA
Judy Day*		Wolf Mayer	ANU	Zbigniew Wasik	NSW
Patrick De Deckker	ANU	Wayne Mayo	ACT	Elizabeth Webber	GA
David Denham	GA	Liz Mayo*		Bruce Webber	ACT
Pat-Ann Denham*		William McKay	GA	Ken Williams	NSW
Fred Doutch	ACT	Kathy McKay*		Lal Williams*	

Geology at ANU (1959–2009)

Bruce Duff	Sydney	David McKirdy	SA	Craig Williams	WA
Jan Duff*		Alison McKirdy*		Neil Williams	GA
David Ellis	ANU	Bear McPhail	ANU	Margaret Williams*	
Anne Felton	ANU	Maria McPhail*		Ian Williams	ANU
Dennis Franklin	Tas.	Louise Miles*		Monica Yeung	ACT
Louise Minty*		Greg Miles	WA	Gavin Young	ANU
Margaret Freeman*		David Moore	Vic.		
Ross Freeman	ACT	Rod Nazer	ACT		
David Gibson	ACT	Jane O'Brien	ACT		
Brian Gibson	ACT	Eva Papp	ACT		
Bruce Goleby	GA	Mervyn Paterson	ANU		
Ann Goleby	GA	Jack Pennington	ACT		
Mark Gordon	Sydney	Peter Percival	GA		
Carmen Gordon*		Paula Percival*			
John Gorter	WA	Brad Pillans	ANU		
Victor Gostin	SA	Sue Pillans*			
Alex Grady	Tas.	Donald Poynton	WA		
Jan Grady*		Richard Price	NZ		

* non-geologist (spouse/partner)
NZ = New Zealand
GA = Geoscience Australia

11
Alumni News

Table 11.1 Contents of Alumni News in date order of graduation

Name	Year	Page	Name	Year	Page
Ian Lambert	1963	174	Doone Wyborn	1983	216
Fred Doutch	1965	175	Michael Andrew		216
Peter Cook	1966	176	Bruce Turner		217
Sue Jephcott		178	Jane Rodgers		217
Martin Carr	1967	178	Jon Olley	1984	218
Peter Lang		179	Robert Creaser		218
David Purcell		181	Oliver Raymond	1985	219
Ian Raine	1968	181	Angela Hume/Thorn		219
Wayne Mayo		182	Paul Johnston		220
Graeme Torr		183	Jyrki Pienmunne		220
David Moore		183	Tim Munson	1986	221
Victor Gostin		183	Simone Veitch		222
Nick Arndt	1969	184	Herman Voorhoeve		223
Wally Bucknell		184	Monica Yeung		223
John Flanagan		185	Kristina Ringwood	1987	223
Anne Felton		185	Steve Sheppard		224
Gavin Young		186	Michael Conan-Davis		224
Bob Day		188	Pauline English	1988	224
Phil Hohnen		188	John Stanner	1990	225
Bruce Nisbet	1970	189	Mark Gordon		225
Richard Price		190	Geoff Deacon		226
Bill McKay		190	Jeremy Peters	1991	226
Andrew Lawrence	1971	191	Dennis Franklin		227
Greg Anderson		191	Cameron Schubert		228
John Wedell		192	Robrt Corkery	1992	228
Gifford O'Hare		193	Chris De-Vitry		229
Ross Clarke		193	Claudia Camarotto		229
John Foden		193	Penny King		230
Dave Gibson		193	Antony Shepherd		230
Don Poynton		193	Chris Pigram	1993	230

173

Name	Year	Page	Name	Year	Page
Tim Griffin		194	Megan James/Spandler		231
John Magee		195	Clinton Rivers		231
Bruce Duff	1972	198	Ron Hackney		231
Rod Nazer		198	Ignacio Martinez	1993	232
John Gorter		198	Tierry Correge		232
David Holloway		199	Roger Skirrow		233
John Brush	1973	200	Greg Miles	1994	234
Phil Creaser		201	Ulrika Troitzsch		235
Gerry Reinson	1974		Tony Meixner	1995	236
John Kennard		201	Tony Rathburn		236
Jim Colwell		202	Tim Barrows		237
Brad Pillans	1974	203	David Tilley	1998	237
David McKirdy	1975	203	Leanne Dancie/Armand	1997	238
Simon Beams	1975	204	Leah Moore		239
Doug Mason	1975	208	Kriton Glenn		239
Mark Stevens	1976	209	Shawn Stanley	1997	239
Ian Smith		209	Dave McPherson	1998	240
Jane Creaser/O'Brian			Sophie O'Dwyer		241
Bill Collins	1977	210	Heather Catchpole	2000	242
Peter Ward	1979	211	Lynda Radke		242
Ray Slater		212	Cameron O'Neil		242
Sue Coote/Slater		211	Michelle Spooner	2001	243
Ian Ferguson	1981	212	Kurt Worden		243
Mukul Bhatia		213	Samantha Williams	2002	243
Joyce Temperly		213	David Murgese	2003	243
Paul Habelko		214	Peter Collett	2006	244
Gordon Taylor	1982	215	Thomas Abraham-James		244

Our graduates have spread across Australia and overseas with employment in many aspects of geology and related disciplines. Some have maintained contacts with each other or the department; others we have lost track of. I have attempted to contact as many as possible by advertisement or word of mouth. Listed here (in graduation date order) are potted histories—some in their own words—of those who responded.

Ian Bruce Lambert

Ian Bruce Lambert was in the first intake of the Geology Department in 1959 (it was then in the Canberra University College), and was its first honours graduate

(1963). After completing a PhD in Geochemistry at the RSES, he was appointed a Research Fellow at the University of Chicago. He returned to Canberra to join the CSIRO, working in the Baas Becking Geobiological Laboratory, which was a joint venture with the Bureau of Mineral Resources (BMR), where he spent 18 years and headed the minerals and stable-isotope programs. After a period as a freelance consultant, he joined the Australian Government in 1990, where he has held a series of senior management positions related to mineral resources, resource access, land use, natural-resource management and environmental protection.

Since 1998, he has led the National Projects, Resources and Advice Group within Geoscience Australia (GA), involving wide-ranging advice, analyses, strategic planning and representational roles in support of minerals policy development and decisions on land use. He is Australian Representative on the Organisation for Economic Cooperation and Development Nuclear Energy Agency/International Atomic Energy Agency OECD-NEA/IAEA Uranium Group (Vice-Chair). He is currently Secretary-General for the thirty-fourth International Geological Congress to be held in Brisbane in August 2012, and an ad-hoc member of the National Committee for Earth Sciences, dealing with international geoscience matters. He represented GA as a Member of the Board of the Cooperative Research Centre for Landscape Evolution and Mineral Exploration (CRC-LEME).

He has published more than 90 refereed scientific research and review papers on a wide range of topics and more than 150 scientific/technical reports. He has made hundreds of formal presentations at major meetings, including more than 50 keynote addresses. Ian has received international research awards, which funded work in: Japan (Japanese Government Award for Foreign Specialists); Europe (Alexander von Humboldt Fellowship); China (Academia Sinica); and Africa (Commonwealth Fellowship). He was also awarded an Australian Government Senior Executive Fellowship in 2000, which funded travel to several continents to study technical advice in relation to sustainable development policy. In 2001, he was honoured to receive the Geology Department's Professor D. A. Brown Medal.

Hadrian Frederick Doutch

Hadrian Frederick (Fred) Doutch was the first mature-age student, graduating with Double Geology in 1965. He had previously started Geology at Sydney University in 1947 together with David Branagan (the future geological historian), and then worked as a geological draughtsman in London and the

Somaliland Protectorate for the British Overseas Survey, and with the BMR in Canberra. He was the first ANU student to study structural geology with Mike Rickard, discussing chapters of the newly published Turner and Weiss text each week. He also included two years of science psychology in his degree—possibly qualifying him as a psycho-geologist!

His early work with the BMR had him looking into local drainage problems, and siting water bores; he graduated to soft-rock and regional field mapping via the Northern Territory and central Queensland to being party leader for a combined BMR–GSQ (Geological Survey of Queensland) survey of the Carpentaria Basin. In the 1970s, he had a major hand in preparing a tectonic map of Australia as a contribution to a Commission for the Tectonic Map of the World project. His draughting experience made him a valuable member of this team. Subsequently, in the 1980s, he was Chairman for the south-west quadrant of a US Geological Survey (USGS) Circum-Pacific Map Project, which overlapped with a two-year stint with the Australian Agency for International Development (AusAID) in the UN Economic and Social Commission for Asia and the Pacific (ESCAP) in Bangkok as an editor/stratigrapher, encouraging member nations to submit basin summaries for publication.

He retired from the BMR in 1986, but later worked off and on as a consultant to its Indonesian–Australian Geological Mapping Project, its jubilee *Geological Map of Australia*, and the aborted Tectonic Map of the Tasman Fold Belt System. Fred is currently eighty-one years old and professes to be moribund, but is mildly bemused by the postulated outcomes of the Larger Hadrian Collector!

Peter Cook

Peter Cook took his BSc (Hons) from the University of Durham (UK). His professional career commenced in 1961 with the BMR, when he worked in central Australia for several years and for one field season in Antarctica. He then took an MSc from the ANU in **1966** on work at the northern edge of the Amadeus Basin.

After two years' research in the western United States for a PhD at the University of Colorado, he returned to work on phosphate deposits in northern Australia. He joined RSES in 1976, where he carried out research into sedimentary-hosted mineral deposits, especially phosphate deposits, and commenced a major study of the palaeogeographic evolution of Australia. His major contribution to the understanding and predictive modelling of phosphorite deposits was recognised in his co-chairmanship of the International Geological Correlation

Project (IGCP 156) between 1977 and 1984. This work led to the publication, in three volumes, by Cambridge University Press of *Phosphate Deposits of the World*. Peter's global perspective continued to be recognised through his subsequent chairmanship of the UN Educational, Scientific and Cultural Organisation (UNESCO) Intergovernmental Oceanic Commission Programme of Ocean Science in relation to Non-Living Resources since 1985. In 1982, he returned to the BMR and, between 1982 and 2000, held a number of senior positions in that organisation including Head of Division and Chief Research Scientist. During this period, he also took a one-year break as professor at the Université Louis Pasteur, Strasbourg, and was a Visiting Fellow at the ANU and the Resource Systems Institute of the University of Hawai'i.

In 1990, Dr Cook was appointed Director of the British Geological Survey. He served on many national and international committees including periods as President of the Association of Geological Surveys of the European Union and the Forum of Directors of European Geological Surveys. He has been a consultant to several governments on reorganisation of their national geological surveys including those of Greece and Finland. He left the British Geological Survey in 1998 and returned to Australia, where he became Executive Director of the Australian Petroleum Cooperative Research Centre (CRC)—an organisation funded by industry, government and academia involving approximately 100 researchers and graduate students throughout Australia. He also continued his involvement with international bodies such as the Intergovernmental Oceanographic Commission, the Advisory Committee on Protection of the Seas, the International Union of Geological Sciences—for which (in 2000) he chaired an international review panel of its activities—and the Academy of Technological Sciences and Engineering as well as the Mineral Industry Research Organisation of Great Britain.

In 1998, he also formed two companies, PJC International Proprietary Limited, which involved him in strategic issues for international mining companies, restructuring issues for a number of geological surveys, scientific reviews and audits, and MineXchange, which has been involved in a number of mineral exploration activities.

At present, he is Chief Executive of the CRC for Greenhouse Gas Technologies (CO_2CRC)—a major research and development initiative focused on the capture and geological storage of carbon dioxide as a greenhouse-gas mitigation option. This brings together more than 100 researchers from universities and research organisations in Australia, New Zealand and internationally as well as many of the world's largest resource companies in order to pursue collaborative research and development.

Dr Cook is a Fellow of the Academy of Technological Sciences and Engineering and has been recognised for his work for science and industry through the award of a CBE, the John Coke Medal, the Public Service Medal, the Centennial Medal, the Lewis G. Weeks Medal and the Leopold von Buch Medal—bestowed by the Deutsche Geologische Gesellschaft during its 156th annual meeting in Leipzig in 2004—honouring his exceptional contributions to economic geology, especially to the geology of phosphate deposits, and his contribution to building and improving national and international geoscientific networks. He has also been awarded a DSc from the University of Durham. He is the author and co-author of more than 120 publications.

Sue Jephcott

'After four years with the Geology Department as a laboratory technician, I completed my part-time BSc in **1966**. I was then fortunate to get a research position with Mount Isa Mines until 1971, when I moved to Tennant Creek as a field geologist with Geopeko for a year. Thereafter, I returned to Victoria to take over the family farm. Collecting all things geological has kept me sane!'

Martin J. Carr

Martin J. Carr was the ANU's first First-Class Honours graduate in Geology (in **1967**), and was then awarded a Commonwealth Postgraduate Scholarship to study Quaternary marginal marine stratigraphy and tectonics at the University of Canterbury, Christchurch, New Zealand, obtaining his PhD in 1970. Marriage and a family—necessitating the earning of a serious living within the mining industry—then intervened in the planned further pursuit of an academic career.

As an exploration geologist, Martin moved from metals exploration in outback Australia, western Canada and southern Africa, to coal in New Zealand, Australia and Indonesia, to gold in New Zealand, Papua New Guinea and west and central Africa. Perhaps his most challenging post was that of Resident Mine Manager at a goldmine in the remote Highlands of Papua New Guinea in the early 1990s, where a wild little man wielding an axe opened with 'Give me a job, or I'll kill you'—a line one does not learn at Harvard Business School! Life as Regional Exploration Manager for a large goldmining company based in Ghana, even stints in the Congo Republic, Burkina Faso, Democratic Republic of the Congo, Senegal and Mali, proved easy going in comparison, and when all else failed in west Africa there was always the option of a cold beer under a coconut palm. Martin finally retired to Tasmania in 2000, relocating in 2007 to be closer to family in New Zealand.

Peter Lang

'Geology set me off on a fantastic career; I would love to be still doing it! My interest in engineering geology started very early. My father was a friend to some of the engineers on the Snowy Mountains Scheme. At an early age, I was taken into the tunnels and onto the dams during construction. At Cranbrook School, Sydney, geology was my favourite subject. My geology teacher, Mr Tebbutt, had a major influence on my career; he enthralled me from second year to the Leaving Certificate in 1964. The ANU Geology Department kept the passion going. The staff were responsible for where I have been and where I am now. In third year, I had a mapping project near Dalgety. My supervisor was Dr Alan White; one day I drove him in my old 1959 VW from Canberra to my mapping area so he could review my work. He never came anywhere with me driving after that! I went on to join the Holden Dealer Team and won the Australian Rally Championship in 1973. I always reckoned a major reason for winning was my ability to read the road conditions ahead due to an understanding of the topography and geology. I also instigated the Canberra Day Car Rally that is now a round of the Pacific Rally Championship.

'I helped build Corin and Googong dams, and all Canberra's sewerage tunnels. David Purcell (ANU Geology, 1969) was my first real boss—at least the first who could influence me. He managed to interest me in a career in geology, as opposed to me using the BMR as simply a source of income to support my rally driving.

'Having solved all the problems with the Googong Dam foundations, and the diversion tunnel (I actually logged some drill core), I went to England and Ireland for a year and met Anne. I came back to Australia, and the BMR. Then one day my boss walked into the office and said 'Anyone want to go to Canada for a year?' BC Hydro and Power Authority were building the Revelstoke Hydro Project. They wanted engineering geologists with real tunnelling experience. Quick as a flash, I put my hand up, and I was off. I worked on the Revelstoke project for two years and learnt an enormous amount. We had to instrument and stabilise a landslide 2 km by 2 km by 250 m thick that had been moving very slowly ever since the ice sheets retreated. If we did not stabilise it, it could slide into the new reservoir, and create a wall of water that would break the

14 dams downstream on the Columbia River, wash away the city of Portland, Washington, and the Hanford nuclear facility, where the nuclear waste from early weapons development was stored in drums on the floodplain!

'From Revelstoke, I went back to London for a year to do my MSc in Engineering Rock Mechanics at Imperial College (and got a son). I returned to Canada to the Syncrude tar sand mine. At the time, it was the largest mine in the world in terms of material moved per day. I learnt lots there, too. Then to Montreal Engineering Company and hydro-electric projects in the Yukon, Newfoundland, Nigeria and Guyana (got a second son). Then to foundation investigations for the Wolsung nuclear power station in Korea. By the way, this power station has the best lifetime capacity factor of any nuclear power station in the world. I say that demonstrates the value of excellent geological investigations (but not everyone can see the connection). Next I managed the rock mechanics program in the Canadian Underground Research Laboratory—part of the Canadian Nuclear-Fuel Waste Management Program. This was the best time of my career. Canada provided what seemed like unlimited funding for scientists to conduct experiments in the Canadian Shield in a dedicated facility we built ourselves at a depth of 400 m. We installed instruments and drilled and blasted and heated and pumped water and injected tracers and measured everything imaginable over distances up to hundreds of kilometres. We worked collaboratively with scientists from the other countries involved in nuclear power and waste disposal. The program was better than climate-change scientists could ever hope for—and far more relevant for the planet!

'It was back to Australia in 1989 to build the very fast train from Sydney to Melbourne via Canberra (and to educate our two boys in Australia). The project fell through, so I did many small jobs for the Snowy Mountains Engineering Corporation. Then I went to Energy Research and Development Corporation, where I managed the Commonwealth's investment (seed funding) in 70 energy research and development projects around Australia—including new coal-power technologies for geosequestration; solar thermal power; methane extraction, liquefaction and powering of coal trucks from Appin to Wollongong; the gas buses in all the capital cities; natural gas-powered intercity road transport; and many energy conservation projects.

'As Tim Fischer's Resources and Energy Adviser, I produced the Coalition's Minerals and Energy Policy for the 1993 election. After the election loss, I joined ACT Electricity and Water (Actew) and worked on the Environmental-Improvement Project at the Lower Molonglo Water-Quality Control Centre. At Actew, I started moving into project management and cost and schedule control systems. I then set up my own business providing consulting advice to various agencies. This was lucrative for five years and has contributed virtually 100 per cent of our superannuation savings. Right now, my position title is "self-appointed expert on everything that matters".'

David Purcell

'I finished study at the ANU at the end of 1967, which is more than 40 years ago, and I cannot believe how time has flown! The only student from that period I have been in regular contact with is Peter Lang, with whom I used to work in the early 1970s.

'My first job after graduation was as an engineering geologist with the Snowy Mountains Engineering Corporation (SMEC). With SMEC, I was involved with feasibility studies and construction of major civil engineering projects such as dams, power stations, tunnels and pipelines. A few years later, I joined the BMR as a senior geologist doing similar work. After 12 years with the BMR, I left to become a consultant in the engineering field. Within a year, I decided on a career change and became involved in gold and gemstone prospecting and, before too long, was mining Queensland boulder opal between Winton and Quilpie. I mined, processed, retailed and wholesaled these wonderful gems at the height of their popularity between 1982 and the late 1990s. With the help of my wife, Louise, we operated a screening plant, D9 dozer, 30 t excavator and drill rig in an open-cut operation covering several leases to depths of up to 20 m. We completed mining and restoration work on our last lease at the end of the 2006 field season and then retired—although I am now working full-time for a short period in my son's IT and brokering business. I have lived on the Gold Coast since 1980.'

Ian Raine

'Mine is a very simple story. After graduating BSc (Honours) in Geology from the ANU in 1968, I worked briefly in Sedimentary Petrology at the BMR during their work on the Sydney Basin. I left the BMR in 1969 to commence a PhD study in the former Department of Biogeography and Geomorphology of the Research School of Pacific Studies, ANU. My thesis, completed in 1974, was on pollen transport and sedimentation and the late Quaternary vegetation history of the Snowy Mountains. I was employed as a palynologist by the New Zealand Geological Survey in Lower Hutt at the end of that year, and I have been there ever since.

'I live in Wellington, and have two adult children: my son, Gareth, is a GIS specialist with the NSW Rural Fire Service in Sydney, and my daughter, Frances, in Canberra assists with care of the elderly. My main professional role in New Zealand has been in applying palynology to dating Permian to Cenozoic strata for regional geological mapping, sedimentary basin studies, and petroleum and coalfield exploration. I have also been involved with development of the New Zealand Fossil Record File and Stratigraphic Lexicon databases, and recently a catalogue of New Zealand fossil pollen and spores. Current projects include taxonomic and biostratigraphic study of New Zealand Jurassic and Cretaceous palynomorphs, vegetation and climate change at the Cretaceous–Tertiary and Paleocene–Eocene boundaries, and biogeography of Cenozoic Antarctic and southern high-latitude floras as a participant in the international Cape Roberts and ANDRILL projects. The Antarctic research entailed laboratory work at McMurdo Station for three summer seasons, and has enabled travel to several overseas conferences in recent years. My partner, Pam, is an archaeologist and I tend to get involved in her projects, too.'

Wayne Mayo

'Graduating in **1968** with a double major in geology and a sub-major in statistics, I first had to complete two years' National Service with the Australian Army. Prior geological fieldwork—including vacation employment with Broken Hill South mapping phosphate deposits at Duchess, Queensland—actually helped the transition to service in the infantry corps. In 1970, after National Service, I joined the then BMR and, while there, working on the geochemistry of the Broad Sound Estuary in Queensland, I completed an economics degree. This degree included a computer science major driven both by the large amount of computing and mathematical/statistical analysis involved in the estuarine research study and by encroaching demands for computer expertise in the geological branch of the BMR.

'Economics beckoned and, in 1976, I joined the Industries Assistance Commission (now Productivity Commission). The transition to economics was made easy by the research culture at the commission (weekly seminars, much statistical analysis, lively discussions during tea breaks, and so on) and the common use with the BMR of the CSIRO's mainframe computer. Direct links to the geological world resurfaced there with involvement in commission inquiries such as the taxation of the mining and petroleum industries and crude-oil pricing. The focus on publishing in the geological world well embedded, my first economics publication (to do with resource-rent taxation) arose out of this work on the latter inquiry. But more formal economics study was required and, in 1979, I completed an honours-equivalent year and joined the Commonwealth Treasury in 1980 to get closer to the policy action.

'During 20 years in the Treasury, I worked on taxation reforms, microeconomic reforms, privatisation of Australian Airlines and Qantas (on secondment to the Department of Finance) and macroeconomic policy issues (including the Treasury macroeconomic model). The odd publication was possible, particularly in the early years, as was some computer modelling. Since leaving Treasury in 2000, I have been "back to the future" in terms of getting up to date with the latest computer programming software for use in such tasks as building a "complete" business income-tax model (in part under contract with the Australian Taxation Office) and associated consulting work.'

Graeme Torr

Graeme Torr graduated with Honours from the ANU in **1968** and an MSc from Imperial College London in Mineral Exploration in 1980. 'I worked with BHP Minerals as exploration geologist from 1969 to 1999, in Australia, Papua New Guinea and Indonesia searching for a variety of minerals; then, from 2002, worked with the Victorian Department of Primary Industries in the petroleum information section.'

David Moore

David Moore claims he struggled through his degree, finally graduating in **1968**. He then worked with Peko-Wallsend, spending a year on fieldwork at Tennant Creek, and, after a break for National Service, another two years at Tennant Creek in the copper–gold mines. He then searched for uranium for the Atomic Energy Commission until the project was shut down for political reasons. From 1975 to 1979, he joined the NT Geological Survey, and then joined BHP in Melbourne working on gold exploration, including two years in New Zealand and as a strategic analyst and interpreter of regional magnetic and gravity data sets. Since 1995, he has been doing similar work with the Geological Survey of Victoria.

Victor Gostin

Victor Gostin studied the lower Permian glaciogene sediments in the Durras–Ulladulla area on the South Coast of New South Wales for his PhD (**1965–68**), supervised by Keith Crook. From 1970 until the present, he has worked at the

University of Adelaide, rising from Lecturer to Senior Lecturer in 1976, to Associate Professor (1994) and Head of Department (1992–95). He retired in 2001, but continues as a Senior Research Fellow.

His research work has been concerned with Precambrian glaciogene sediments, regolith and landscape development, and the effects of meteorite impacts (the Acraman Impact) in the ancient sedimentary record. He has edited or contributed to two books and numerous journal and conference articles. He has broadcast regularly on Adelaide radio, and lectured periodically to schools and clubs on geoscience and environmental issues.

Nick Arndt

'I have moved around a lot but now am settled in France; it comes of having a French wife, Catherine Chauvel, who is also a geochemist, and the French do not transplant well! Since graduating from the ANU in 1969, I worked for a year in a fly-by-night company during the first nickel boom, did a PhD in Toronto (1975) and postdoctorates at the Carnegie Institute in Washington, DC, and at the University of Montreal. I also taught for three years in Saskatoon before reverting to postdoctoral status, subsequently upgraded to that of Wissenschaftlicher Mitarbeiter, at the Max Planck Institute in Mainz, Germany. I then accepted a research–teaching position in the University of Rennes and now another position at the University of Grenoble, where I have the exalted position of "Professeur classe exceptionnelle echelon 2". My research is mainly along two lines. The first involves komatiites and, more generally, the conditions at the surface and in the interior of the Archean Earth; the second deals with large igneous provinces, their origin, their ore deposits and their impact on global climate.'

Wally Bucknell

'After graduating in 1969, I went to Canada with Peter Moignard and explored for uranium in northern Saskatchewan for two years, and then spent 11 years with Rio Algom in the Canadian Shield and Appalachians (VMS, gold, uranium, potash, SEDEX, whatever). I then moved back to Perth in 1984 to join Noranda Australia, which became Plutonic Resources, and that was a real buzz through the 1990s, with lots of (gold) discoveries and big exploration budgets. Since Plutonic was taken over about 10 years ago, I have been running an ASX junior (Atlantic Gold), whose main asset now is an emerging open-pit gold development in Nova Scotia, which should be in production in about 18 months.

'I live in Sydney with my wife and three (more or less grown up) kids: oldest son at IBM, middle daughter a nurse and living at home, and youngest son also at home and finishing Engineering/Science at Sydney University. Note, no geologists!'

John Flanagan

'I finished my degree in **1969**, majoring in Geology, whilst on a scholarship for the NSW Department of Education. The next year, I went to Sydney University and completed the Diploma in Education. In 1971, I started to work as a science teacher for NSW Government schools and worked for the Education Department for my entire working life, almost always in the NSW country, and I have now retired in Orange. The Geology degree I completed at the ANU was very comprehensive. My eldest son followed my interest and became a mining engineer. When I graduated, it was the height of the Poseidon nickel boom and now it is the iron-ore boom more than 30 years later.'

Anne Felton

'Planning a career as an industrial chemist, I embarked on first year at the ANU—my "local" university. Not wanting to do a biological subject, I took Geology A01 in first year as my fourth subject (in addition to Chemistry 1, General Maths and Physics 1). Geology opened up a new, wonderful world of crystals, fossilised remains, folds and faults, and 100 different minerals to learn in first year alone. The field trips, however, with Geology Department staff to explain how rocks worked and soils formed made sense of the lectures and theory, and the prospect of having the whole world for a laboratory instead of just a room in a building won me over. In third year, I did a double major in geology.

'I began first year in 1964 and graduated with Honours in **1969**, having gone part-time for a couple of years to repeat failed subjects (Chemistry) and earn money. At the end of third year, I was awarded a NSW Department of Mines Cadetship, so I could afford to stay on at university for another year! Thus, I became the first woman to complete the four-year Honours Degree in Geology at the ANU. This launched me on a geological career that I have stayed with ever since, and most of the projects I have worked on have been field based.

'My clearest memories of my time as an undergraduate in the Geology Department are of the inspiring presence of Professor David Brown, and the very high standards of teaching. Our lecturers were passionate about their specialty subjects and were always willing to explain difficult concepts (often several times). The enthusiasm of the academic staff for both their teaching and their research, and excellent support from the technical staff, made for a lively department. The excellent teaching and emphasis on field studies have proved to be very good training for the variety of geological work I have done with the NSW Geological Survey and the (then) Bureau of Mineral Resources, Geology and Geophysics. Even later, I was encouraged by several of my former teachers to undertake a PhD degree in sedimentology at the University of Wollongong, which I completed in 1993.

'I returned to the Geology Department as a Visiting Fellow in 1992, and again in 2005 after a period of research at the University of Hawai'i. This research (which I am continuing here in Australia) is about the sedimentology of rocky shorelines and has ranged through depositional processes and environments of the gravelly erosional products that accumulate on them, the impacts of tsunami and storm waves ('Big Rocks'), and the origins of Pleistocene coastal gravels on oceanic hotspot islands. I am writing a series of papers on this work (four published; one in press).'

Gavin Young

'I was introduced to geology in 1966, witnessing Tony Eggleton's first lectures for Geology A01. In the following years, I benefited from a Commonwealth Government Free Place (as a clerk in the Navy Department!) to do full-time university (second and third-year Geology). Instead of work experience as a trainee geologist during vacations, I was assigned some fascinating tasks by the Navy Department, of which the annual furniture inventory is still fixed in my memory. Because I would be bonded to work for the Government on completion of the degree, it was suggested that I also do statistics/economics as well as geology, to give some career options in the Public Service other than the BMR, which was not highly regarded in some quarters. This advice was not followed exactly!

'During the big drought of 1968, Geology Department staff (including Professor and Mrs Brown and Dr Campbell) had visited Cave Island, Lake Burrinjuck, to examine a tract of well-exposed limestones that had been underwater for years. They discovered abundant fossil bones of Devonian fish, and this was the material researched for my honours thesis, combined with mapping of the limestones in the Wee Jasper Valley, for which I got First-Class Honours in **1969**.

'My geological employment began in 1970 in the BMR museum, with responsibility for sorting out the palaeontology collections, but with 50 per cent of the time allocated for researching Devonian fossil fish. A significant opportunity arose to visit Antarctica with Dr Alex Ritchie of the Australian Museum, to collect fossil fish in the Transantarctic Mountains. With BMR support from John Casey (Assistant Director), I joined the fifteenth Victoria University of Wellington Antarctic Expedition (VUWAE 15) led by Dr Peter Barrett (Victoria University). The 1969–70 VUWAE had discovered abundant fossil fish in the Devonian Aztec siltstone. Many of the localities were within the eastern sector of the Australian Antarctic Territory, but completely remote and inaccessible from Australian Antarctic bases. The New Zealand Antarctic Research Program operated from Scott Base (with US logistical support). It seems that the only 'public servant' from Canberra previously sent to this region to 'geologise' in the employment of the Commonwealth Government was T. Griffith Taylor, who participated in Scott's British Antarctic expedition to the South Pole in 1911–12!

'Living for three months in tents and travelling by toboggan and sled in the Transantarctic Mountains during the summer of 1970–71 were memorable field experiences. It also set the scene for many decades of research on the Aztec Devonian fish fauna (nearly finished in 2008). Some 433 published pages in journal articles and monographs (a few with co-authors Carole Burrow, John Long, Alex Ritchie and Sue Turner) have documented nearly 50 genera and species, making the Aztec assemblage one of the most diverse Devonian fish occurrences known throughout the world for the middle–late Devonian.

'An Australian Government scholarship permitted me to go to London to study for my PhD in the British Museum (Natural History), which held some 500 fossil fish specimens from the Burrinjuck area collected during two British expeditions to Australia in 1955 and 1963. Some exceptional skulls and braincases collected from Wee Jasper during my honours year formed the basis for a PhD in Zoology at the University of London, in 1976.

'The next 20 years of research with the BMR (which became AGSO, then GA) focused increasingly on using fossils to solve geological problems, with fieldwork in the Canning Basin (1972), Amadeus Basin (1973, 1984, 1991), Georgina Basin (1974, 1977), and many localities in south-eastern Australia, including geological mapping of the upper Devonian Hervey Group on the Forbes 1:250 000 sheet under the "National Geoscience Mapping Accord" during 1994–96. Overseas conferences and fieldwork included New Zealand (1980), China (1981, 1987, 1992), the United States (1985, 1992), England (1988), Venezuela (1992),

Malaysia (1993), France and Germany (1993, 1995), and Vietnam (1995)—all externally funded except the BMR-funded US trip in 1985. During this period, I was involved in organising two significant conferences in the Geology Department—the first with Ken Campbell (ANU) and Alex Ritchie (Australian Museum) being the International Symposium on Early Vertebrates held in Sydney and Canberra in February 1983 (see Plate 5.1); and, a decade later, when President of the Association of Australasian Palaeontologists, a symposium was held to honour Professor Ken Campbell on his retirement from the ANU staff (February 1993). Both resulted in published symposium volumes.

'As the union leader in the Australian Geological Survey Organisation (AGSO) during 1996, when "downsizing" of the Public Service under the Howard government impacted on geologist positions, I had numerous meetings with fellow ex-ANU Geology student Neil Williams (by then AGSO Director). A committee of 13 was formed to fight the proposed retrenchments of many geologists. The success of this endeavour is indicated by the fact that (along with many other geologists) all 13 committee members were themselves retrenched! I subsequently became a Visiting Fellow with DEMS.'

Bob Day

Bob Day completed his PhD in Palaeontology in **1969** as a staff candidate. He subsequently worked for Esso Exploration and Production Proprietary Limited for three years as a petroleum geologist. He then joined the Geological Survey of Queensland in 1973 and held several positions: Principal Geologist, Assistant Chief Government Geologist, Director of the Fossil Fuels Branch, and Director of Regional Investigations. Then, in 1986, he took an MBA and became Principal Coordinator and then Manager of Information Services for the Department of Mines and Energy. In 1988, he went back to the Geological Survey as Chief Government Geologist. In 1995, he took early retirement, but, after six months, returned as Director-General of Mines and Energy, then Chairman of the Coal Board and Electricity Regulator. In 1998, he retired to become a Director of Arrow Energy, and, in 2006, non-Executive Chairman of Pure Energy Resources Limited.

Phil Honen

'Isn't the Net wonderful? Here I am 40 years on, in northern Thailand, and you track me down! Well done, and good to hear from you. After graduating in **1969** and working for the BMR, I had two successful stints at opal mining—the first (1977–81) ended when we got greedy and split into opal and gold-prospecting teams. David Purcell moved our operations north and found a lot of boulder opal that lifted off the ironstone substrate and hence would not cut into stones. I then prospected for alluvial gold along 80 km of the Palmer River on Cape York Peninsula. We found interesting prospects, but the bottom line was our report was delayed in reaching the Queensland Mines Department, and they would not renew our Authority to Prospect, despite a meeting with the Minister for Mines.

'I decided that it was all too hard to return to geology after working at what some regarded as a non-professional field. I had a home by now on Queensland's Gold Coast and worked at a variety of jobs before being asked, in 1987, to return to opal mining as a consultant. What transpired, briefly, is that I found, quickly, millions of dollars worth of very high-quality boulder opal, but was cheated out of my share while on a break.

'I became totally disillusioned and decided to stay home with my family and opened an antiques business, which I stayed with until completing Bible College in 1997. After that I visited the United Kingdom and Israel and helped Creation Ministries make some videos. After this, I travelled to Papua New Guinea on mission work and finally to Thailand in March 2000. I came to Chiang Mai seven years ago at fifty-four years young to start a children's home for abandoned and HIV-positive babies, and a new life. Five years later, we have 18 staff, 24 children ranging down from five-plus years—all healthy and being educated inhouse on 3 acres (1.2 ha) of fruit trees on a terrace overlooking the Ping River at a rural location called 'Hot', 111 km south of Chiang Mai. We started with bare soil and now have a school, three dormitories, a kitchen/dining Bali-styled hut and a new recreation yurt, as well as teachers' cottage and laundry building. We still need to set up our own web site, but others mention us—"Baan Fah Sighy" or, translated, "Blue Skies Children's Home".'

Bruce Nisbet

Bruce Nisbet came from James Cook University of North Queensland (JCUNQ) to take Honours in **1970**. He mapped at Captains Flat and, after graduating, he completed a PhD in Structural Geology at Albany (New York State). He then went enthusiastically into exploration geology with Craig Williams (a fellow ANU graduate)—first with Hunter Resources, and that led to the discovery, among other successes, of the Ernest Henry deposit in 1990. This success was awarded AMEC's 'Prospector of the Year' award. They then very successfully developed Equinox Resources, working from England in Zambia and Peru. Laurie Curtis (another ANU graduate) had just joined Bruce in a new exploration venture in South America when sadly he died of cancer in May 2006. Craig Williams is now CEO of Equinox, and they run the biggest copper mine in Africa.

Richard Price

Richard Price took his Honours Degree in **1970**. He comments that the degree covered a wide range of sub-disciplines and prepared him well for employment in exploration and research careers: 'we learnt most of our geology during field excursions, mapping camps and field projects.' He won an Australasian Institute of Mining and Metallurgy (AusIMM) bursary in 1968, and worked as a research assistant with Bruce Chappell in 1971.

After graduating, he spent a year with MIM at Mount Isa, and then became a Teaching Fellow at the University of Otago, New Zealand, working on the geochemistry of the Miocene Dunedin volcanics for a PhD—gained in 1974. From 1974 to 1998, he progressed from Lecturer to Reader at La Trobe University, also serving as Chair of the Department, Head of the School of Earth Sciences and Dean of the School of Physical Sciences. From 1998 to the present, he has been Professor and Dean of the School of Science and Engineering at the University of Waikato, New Zealand. He has held visiting appointments at MIT (1981–82), University of Auckland (1989), RSES (1989, 1994), University of Melbourne (1997), University of Tasmania (2001), and University College Cork (2005).

W. (Bill) J. McKay

W. (Bill) J. McKay was awarded BSc (Geology, Hons 2A) in **1970** and, some 20 years later, returned to the ANU part-time and completed a PhD in Economic Geology. On completing honours, he joined Jododex Australia as Exploration Geologist based at the Woodlawn base-metal prospect in New South Wales. He was involved in exploration and feasibility investigations of the Woodlawn deposit until 1974, and then joined Mount Isa Mines as Senior Research Petrologist, Milling Research, based in Mount Isa. He returned to Canberra to work at Woodlawn in the late 1970s as Senior Mine Geologist (later Chief Geologist, then GM Resource Development) for the Woodlawn Mines Joint Venture. The 1980s was an interesting and challenging period as the joint venture wrestled with low metal prices, difficult massive sulfide (complex ore) metallurgy and transition from open-cut to underground mining at Woodlawn.

In the late 1980s, he joined the Australian Public Service (BMR), primarily to remain in Canberra and write up his PhD. During the early 1990s, he was seconded to the Department of Foreign Affairs and Trade to provide technical advice on matters relating to deep-seabed mineral exploration and mining under the UN Law of the Sea Convention, prior to Australia becoming a signatory to and then ratifying the convention. From this, he gravitated into numerous and varied technical advice/assessment roles for the Australian Government, while employed by the Bureau of Resource Sciences and AGSO. These included a temporary (six-month) secondment to the Northern Territory as Director of the Geological Survey and assignments advising on institutional strengthening in Fiji and addressing issues relating to the exploration and mining of phosphate on Christmas Island and Nauru. Currently, he is employed as Executive Officer, Onshore Energy and Minerals Division, at Geoscience Australia (GA).

In 1970, he married Cathy O'Donnell (BSc Hons Zoology ANU and BEc University of Queensland). They have a daughter, Valerie (solicitor, living in London), born in Mount Isa, and sons Kenneth (deceased) and Owen (studying medicine at Newcastle), born in Canberra. Valerie and her husband, Michael, presented us with our first grandson, Daniel, in March 2008.

The information on the following group of five students who graduated in the early 1970s was submitted by Andrew Lawrence.

Andrew Lawrence

From 1981 to 1983, Andrew Lawrence worked in coastal Saudi Arabia, Sinai and Yemen for Seltrust Engineering searching for evaporates, especially potash. He also worked in Iran and Liberia. He currently manages the Perth geological section for Hatch—a large global consulting and engineering company.

Greg Anderson

After graduation, Greg Anderson went to Papua New Guinea as an engineering geologist. He is currently Executive Director for the PNG Chamber of Mines and Petroleum. He gives presentations on development prospects in Papua New Guinea at various forums in Australia.

Geology at ANU (1959–2009)

Plate 11.1 The Grogans in 1974: Gifford O'Hare, Greg Anderson, John Wedell, Ross Clarke and Andrew Lawrence
Photo from Andy Lawrence

Plate 11.2 The Grogans in 2008: John Weddell, Greg Anderson, Andrew Lawrence, Gifford O'Hare, Ross Clarke framed
Photo from Andy Lawrence

John Wedell

John Wedell has now retired from teaching and various IT roles and lives in Sydney.

Gifford O'Hare

Gifford O'Hare is currently teaching in Sydney.

Ross Clarke

Ross Clarke graduated in **1971**. He was President of the Student Geological Society and organised help after the road accident at Gundagai wrecked one of our homecoming field vehicles. He made his career in Johannesburg, South Africa, where he is now General Manager for Alfred H. Knight—a cargo inspectorate that specialises in ores, concentrates and metals.

John Foden

John Foden (BSc ANU **1971**; Hons Tas. 1973; PhD Tas. 1980) worked as exploration geologist with Esso (1972–74); Research Assistant, Macquarie University (1979–81); and postdoctorally at Adelaide University (1981–84), and then was a staff member, rising to Professor in 2005. He has served as Head of Department, as President of the Geological Society of Australia (GSA), and on the editorial boards of several journals. His petrological and geochemical research has concentrated on magmas in modern and ancient island arcs, especially in Indonesia, and on the origin of granites.

Dave Gibson

Dave Gibson took his Honours at ANU in **1971**, specialising in soft-rock geology. Then, from 1972 to 1985, he worked with the BMR on regional mapping in the Carpentaria and Canning basins, and spent two years in Papua New Guinea and several years researching oil-shale occurrences. In 1983, he became one of the founding members of the BMR Regolith Group. After 1985, he left geology for a life of leisure on the NSW South Coast—lawn mowing, teaching music and wholesaling florists' greenery. Returning to Canberra in 1995, he took up contract work in regolith and geomorphology for the CRC-LEME based at GA.

Don Poynton

'Since graduating in 1971, I have been fortunate enough to have never had a day without a job—somewhat unusual in the very cyclical petroleum exploration business. Over the past 35-plus years, I have worked for Geophysical Services

International (1972), BOCAL/Woodside (1972–79), Mesa (1979–84), WMC Resources (1984–97) and, since 1997, for a small Australian explorer and producer, Strike Oil. These days, I am fortunate to be able to combine two of my interests and wear the hat of Manager of Exploration and Environment. A majority of my work has been in Australia, although two years in the world's oil capital, Houston, with WMC, were most rewarding. Probably my best memories, however, are of a year spent on the island of Timor in 1974–75 doing geological fieldwork and follow-up drilling. We were aware that Indonesia was going to invade Portuguese Timor, as our calls for a doctor after an aircraft accident were met with 'no doctors available, all been sent to border', and the next night their warships passed close to the island we were working on. We left Timor a few days before the invasion.

'As most of my time is now spent in the office, I try to incorporate a bit of outdoors geology into my holidays. Some of my more adventurous tours have included Patagonia, Antarctica and Mongolia. To keep my wife happy, we also embark on some more idyllic walking holidays such as following the footsteps of William Smith (in Somerset)—inspired by the map that hung in the old lecture theatre—exploring the Dinosaur Coast (in Yorkshire) to view the cliffs where biostratigraphy was born and the Jurassic Coast (in Dorset and Devon) to view the wonderful exposures, and, of course, look for the odd fossil or two.

Tim Griffin

Tim Griffin completed his honours project on the Moruya granite under Bruce Chappell in December **1971**—the year after the nickel boom collapsed. 'I then joined Alan White at La Trobe University in 1972 as the second staff member of the brand new Geology Department to prepare first-year practical class materials. This was lots of fun, and included weekend field trips to get samples for rock and mineral sets. Whitey had just returned from sabbatical in the United States and had a great American accent and cowboy swagger to stir up the first-year students. I married Steph Day, an ANU Biology graduate who also did first-year geology. We arrived in Townsville in the wake of Cyclone Althea for a PhD on the Tertiary McBride volcanics to map lava craters, cinder cones, long lava flows (up to 160 km Undara flow), lava tunnels/tubes and collapses, and so on. I worked on the geochemistry and high-pressure rock and mineral inclusions.

'The industry was still depressed so the odd short visits to Papua New Guinea in 1974 and 1975 sowed the seeds for a move to join the PNG Geological Survey in 1976 to start 1:100 000 geological mapping. I worked around Ok Tedi in the

Star Mountains on the border with Iran Jaya (West Papua), central Papua New Guinea and Mount Wilhelm, and made brief visits to many other parts of the country, including Bougainville, Rabaul and the Sepik.

'With two young children in tow, I joined the Geological Survey of Western Australia and moved to the Kalgoorlie office in 1980 at the start of the gold boom. We began the systematic 1:100 000-scale mapping in Western Australia, where we recently reached our 150th map sheet with a further 450 to go—but only if there is value in mapping a massive area of sand dunes and not much else! We mapped across the eastern Goldfields and the Kimberley as well as reporting on the geology, tectonics and mineralisation across Western Australia as part of a fantastic team of very enthusiastic colleagues.

'I took on a senior management role in the Geological Survey of Western Australia in 1995, as we moved entirely to a digital approach to map production and management of data and information, and watched the massive impact of SHRIMP as it provided age constraints on the Precambrian rock history of Western Australia.

'I have been Executive Director of the Geological Survey of Western Australia for seven years now and have seen the growth in budgets and staff—and many changes in the approach to the mapping, capture, storage and delivery of geoscience information; but the underlying fundamentals of geology remain. Dealing with this change and the political issues—particularly the competing interests that impact on land access for exploration—provides special challenges, and it is not pleasing to see the ignorance within the general community of the fundamental role that geology and resources plays in our day-to-day lives. What is exciting is the work in Western Australia on the evolution of the early crust, atmosphere and life on Earth, and the crossover between research in molecular biology and geochemistry. It is very pleasing to see the value that the geoscience fraternity continues to place on good regional geology studies, in all its forms, as commodity prices rise, geological models evolve in the search for ore bodies with no surface expression, and as new applications appear, such as carbon-dioxide geosequestration and geothermal energy.'

John Magee

'My direct connection with the Department of Geology extends for 40 years—from 1968 to 2008. I completed a BSc pass degree with a major in Geology, plus some Zoology and Physical Geography sub-majors in **1971**. I had a very undistinguished undergraduate record and did not gain entry to honours, which was a good thing as I had not found a specific focus for my geological interest, and would have done badly. I did not think so at the time, and was very disappointed. I then managed to contract a major illness, which laid me low for some 18 months till late 1972 when I recovered enough to seek employment, without any success. Then in early 1973, when I was about to depart for New Zealand for an extended bit of travelling, a job was advertised in the Department of Biogeography and Geomorphology, RSPAS, ANU, to assist Jim Bowler in working on the Quaternary geology and geomorphology at Lake Mungo, western New South Wales, as well as assisting geological interpretation of archaeological excavations. With a favourable reference from the department, I got the job. This started out as a part-time position and gradually evolved into a full-time role as Jim Bowler's research assistant. I found that I loved the desert, loved working in the Quaternary and Jim was an excellent and inspiring person to work for. The work was interesting and challenging and I thrived. Between contracts, I had a nine-month stint with the NSW National Parks and Wildlife Service doing a resource survey of the Lake Mungo region, which eventually led to the establishment of the national park at Lake Mungo. Nevertheless, I was not happy in the bureaucracy and I was very pleased to return to the ANU working for Jim Bowler in 1977.

'I stayed at Biogeography and Geomorphology, and later, the Division of Archaeology and Natural History, which grew from a merger of the Prehistory and Biogeography and Geomorphology departments, till 1997. The research work evolved away from Lake Mungo and into the desert proper with funding by the ANU of a major interdisciplinary project jointly conducted by Biogeography and Geomorphology and RSES, called SLEADS (Salt Lakes Evaporites and Aeolian Deposits). This project developed specialist equipment to drive on playas and salt lakes and to drill the sediments to reconstruct arid-zone palaeo-environmental and palaeoclimatic history. At the same time, I resurrected my academic career and enrolled part-time in an MSc qualifying course back at the Geology Department, supervised by Keith Crook. Progress was slow with work commitments (often involving long field trips to central Australian salt lakes to supervise the SLEADS drilling crew). In 1988, however, I eventually graduated with an MSc by thesis on "Chemical and clastic sediments and late Quaternary history, Prungle Lakes, western NSW". In July 1982, I led the first SLEADS drilling trip to Lake Eyre, which initiated a long research connection with that lake. After Jim Bowler departed the ANU in the late 1980s and the SLEADS project had finished its fieldwork stage, I stayed on at Biogeography and Geomorphology, and then Archaeology and Natural History, continuing analysis of the SLEADS cores and working on other projects. My main research interest, however, was a part-time PhD on Lake Eyre, which I commenced in late 1989, and the project was enhanced soon after starting when luminescence dating became available, which allowed Quaternary sediments to be dated well beyond the 35–40 000-year limit of radiocarbon. My thesis project title was "Late Quaternary environments and palaeohydrology of Lake Eyre, arid central Australia".

'The period 1996–97 was momentous in my life, as I was made redundant when the geological part of Archaeology and Natural History was abolished by an anti-science RSPAS Director aided by the misguided economic-rationalist regime that flourished at the ANU at the time. I was awarded an ARC Postdoctoral Fellowship in late October 1997, provided my thesis was submitted before the end of the year. That was achieved—about three hours before the ANU closed for its Christmas break. I duly started a new life as an academic back in the Department of Geology in January 1998, working on extending the Quaternary and climate history record from Lake Eyre and comparing it with other sites. Throughout my Lake Eyre research work, I had collaborated with Professor Gifford Miller of the University of Colorado to date sediments by amino-acid racemisation (AAR), U-series and radiocarbon dating of the eggshell of the large flightless ground birds the emu and the now extinct Genyornis. Eggshell of both birds is the most common biomineral in the arid zone and preserves its original chemical composition with great fidelity, which makes it ideal for chronological and isotopic analyses. Almost inadvertently, we managed to date the extinction of Genyornis—an element of the Australian mega-fauna, which all became extinct across the continent about 46–50 000 years ago, soon after the arrival of humans on the continent. The cause of the extinction has been debated widely for some 150 years and our geochronological research and subsequent detailed palaeodietary comparisons of emu and Genyornis, from stable carbon isotopes, have provided valuable new information for resolving that mystery. Tim Flannery described this research as having "broken the deadlock" in the mega-faunal extinction debate. Work on Quaternary palaeo-environments and mega-faunal extinction continued into an ARC QE 2 Fellowship until 2006, and was an exciting and productive episode.

'At the end of the QE2 Fellowship, however, I was unable to obtain further ARC fellowship support and research activities have come to a disappointing end, with many directions unfinished or barely started. I now find myself, like many geologists probably do, unemployed and without any realistic avenues of employment. I have had a few contracts at GA and the Bureau of Rural Sciences, but I am too old to reinvent myself and too young (and with insufficient and dispersed superannuation) to retire. I have been a Visiting Fellow in Earth and Marine Sciences and now RSES since the end of my ARC positions.'

Bruce Duff

Bruce Duff graduated in **1972** and then worked for the BMR during 1973 and 1974, before gaining his PhD in Palaeomagnetism from Leeds University in 1978. After working as a research assistant at the ANU on the Wagga-to-Batemans Bay structural transect during 1979–80, he was employed by Shell International as a geophysicist—in the Netherlands from 1980 to 1984, then in Oman (1984–87). Several international postings with the Belgian company PetroFina followed (1987–99)—first as an explorationist, then as a negotiator within their commercial department. Total's acquisition of PetroFina in 1999 led to a further three years of similar commercial work. In 2003, Bruce returned to Australia and joined Oil Search as New Business Manager.

Rod Nazer

Rod Nazer was a keen student in the ANU Geology Department from 1967 until **1972**, completing his honours year under the supervision of Professor Ken Campbell. He then moved to Canada and commenced postgraduate studies at the University of Toronto. Owing to the transfer of his supervisor to the University of Queensland, Rod moved to Brisbane in 1974, and graduated in 1977. He then took a position at Canberra Grammar School (CGS) as a specialist geology teacher, completing a Diploma in Education (with Distinction) through Mitchell College, Bathurst, in 1981. After 10 years at CGS, Rod took a position as Science Coordinator at Geelong Grammar's Timbertop campus. Returning to Canberra in 1990, Rod taught for 11 years at St Edmund's College—the last eight years as Science Coordinator. He returned to CGS in 2001 and enjoys teaching science (including geology) to students in years 7–10 and physics to HSC students.

John D. Gorter

John D. Gorter graduated in **1972** with Honours in Marsupial Palaeontology, and, in 1993, with a PhD from the University of New South Wales on the Cambro–Ordovician siliciclastic reservoirs and source rocks in the Amadeus Basin. From 1972 to 1977, he worked for the BMR—first as a field hand, then as a petroleum geologist; from 1978 to mid-1980, he was with ESSO Australia in Sydney, with six months in Canada with Imperial Oil; then from 1980 to 1986, he worked with Pancontinental Petroleum Limited and then with AGIP Australia drilling exploration wells in Queensland to mid-1987, followed by three years consulting. Since 1990, he has worked out of Perth—first with Petroz NL; then, from 1993 to 1999, with Hardy Petroleum Limited, with discoveries including Bayu (approximately one billion barrels of oil equivalent), and the Woollybutt oil field; from 1999 to 2000, with British-Borneo Australia Limited; and since 2000, for Eni Australia (currently New Ventures Manager) exploring the North-West Shelf and Timor Sea.

He has authored more than 50 refereed technical papers and articles in trade journals and has made many technical presentations, with best paper award from the Australian Petroleum Exploration Association in 1984. He ran three successful 'Western Australian Basins' symposia, including one as chairman. He won a Meritorious Service Award from the Petroleum Exploration Society of Australia (PESA) in 1999, and co-wrote and performed in two 'rock operas' for PESA Christmas events. He has been married for 30 years to Eva Skira, and has two daughters (aged twenty-one and seventeen) and a son (aged fifteen).

David Holloway

'After completing the honours year in **1972**, I worked in the Geology Department for several months as Research Assistant to Ken Campbell before commencing PhD studies late in 1973 at the University of Edinburgh, under the supervision of Euan N. K. Clarkson, on the trilobite fauna of the St Clair limestone (Silurian) in Arkansas. My study of this fauna utilised the extensive collections of the US National Museum of Natural History, Smithsonian Institution, which I visited in 1974 and again in 1976, and during my first visit specimens were also collected in the field in northern Arkansas. I completed the PhD in 1977, and, at the beginning of 1978, returned to Australia and to the ANU Geology Department—initially as Research Assistant to Ken Campbell and later as Tutor. In September 1979, I was appointed Assistant Curator in Invertebrate Palaeontology at the National Museum of Victoria (now Museum Victoria), Melbourne, later becoming Curator, then Senior Curator and, most recently, Deputy Head of Sciences (Geosciences).

'My research has focused on the Silurian and early Devonian trilobites of eastern Australia, in the course of which I have collected extensively in north Queensland, central western New South Wales, Victoria and Tasmania. I have also been undertaking taxonomic revision of certain trilobite groups for the updated trilobite volumes of the Treatise on Invertebrate Paleontology, work for which is still in progress. In 1983–84, I spent 12 months at the University of Aston in Birmingham, UK, on a Leverhulme Fellowship, working on a revision of lichid trilobites with Alan Thomas. In 1990, I visited the Soviet Union under the Australia–USSR Bilateral Science and Technology Program, in order to study trilobite collections in the Paleontological Institute, Moscow; the CNIGR Museum, St Petersburg; and Tartu University, Estonia. In 1993, I was awarded a German Academic Exchange Service (DAAD) 'Study Visit' grant to examine trilobite collections at the Museum für Naturkunde, Berlin, the University of Göttingen, the University of Bonn, the Senckenberg Museum, Frankfurt am Main, the National Museum of Prague, and the Czech Geological Survey. In 2001, I again studied trilobite collections, in Germany and the Czech Republic, and in the Natural History Museum in London.

'Conferences I have attended recently include the third International Conference on Trilobites and their Relatives, Oxford, 2001; the tenth International Symposium on the Ordovician System and the third International Symposium on the Silurian System, Nanjing, 2007; and the fourth International Trilobite Conference, Toledo, Spain, 2008. I have worked with a number of other scientists both from Australia and from overseas, including Ken Campbell, Peter Jell (at that time at Museum Victoria), Fons Vandenberg (Geological Survey of Victoria), Phil Lane (Keele University, UK), Alan Thomas (now at the University of Birmingham, UK), Barrie Rickards (University of Cambridge), and Maria da Gloria Pires de Carvalho (American Museum of Natural History). I have jointly supervised one student, Andrew Sandford, working on Silurian and early Devonian trilobites for the MSc and PhD degrees at the University of Melbourne.

'I have served on the committee of the Victorian Division of the Geological Society of Australia, and was Editor of the Proceedings of the Royal Society of Victoria from 1989 to 1993. In 2004, I was elected as a titular member of the International Subcommission on Silurian Stratigraphy.'

John Brush

'After I left the ANU at the end of 1973, I joined the BMR as a field hand on a regional mapping party. Not long after, I took up a "short-term" position in Canberra as a graduate in the Commonwealth Public Service, where, much to my amazement, I remained for the next 32 years. Most of my public service career was devoted to a range of resources policy and program areas, with a focus on nickel and coal, facilitation of major resources projects and petroleum. In my most recent position, I managed Australia's offshore petroleum exploration areas. I retired in 2006 so that I could devote more time to my major interests of travel, caving, skiing and walking.'

Phil Creaser

Phil Creaser (BSc Hons **1973** and MSc 1977) worked in various Commonwealth environment and heritage departments in Canberra for most of the time from late 1973 until he retired in 2006. While some positions enabled him to utilise his geological knowledge, other jobs were free of any geology content. Nevertheless, he always maintained his interest in geology—particularly in geological heritage issues—and was elected to the International Task Force on Earth Heritage Conservation following the Malvern Conference in 1993. He has worked with Professor Michael Archer and his palaeontological team at the University of New South Wales since the mid-1980s and played a key role in the inscription of Riversleigh on the World Heritage List in 1994. He established the CREATE Foundation at the University of New South Wales, where he is currently a Visiting Fellow. His retirement has allowed him more time for his broad geological interests and recently he has been on trips to New Zealand to work on Miocene fossil sites and to Cape York to collect amber. He also spends a lot of time orienteering and rogaining, with some successes at both the national and the international levels, and, in between these interests, he tries to attend as many trivia nights as time permits.

John Kennard

John Kennard obtained a First-Class BSc Geology (Hons) in **1974** and was awarded a University Medal—the first Geology graduate at the ANU to receive that award. 'After graduating from the ANU, I travelled, before commencing work with the BMR in May 1975. I am still there, but have gone and come back a couple of times! My first stint at the BMR was 1975–79, working as a sedimentologist and mapping the Cambrian carbonate successions in the Georgina and Amadeus basins, as well as some great field studies on the Great Barrier Reef. I left in late 1979 and worked in New Zealand at Lincoln College, Christchurch, working on late Quaternary deposits and geomorphic analysis of land systems.

'I returned to Australia in 1981, and recommenced with the BMR working on the sedimentology and petroleum geology of the Amadeus Basin. I was awarded an overseas Commonwealth Postgraduate Scholarship in 1983 to undertake a PhD in carbonate sedimentology at Memorial University, Newfoundland, Canada. After two years' full-time study in Newfoundland (1983–85), I returned to Australia and the BMR, completing studies on a part-time basis before receiving my PhD in 1989 (my thesis was titled 'The structure and origin of Cambro–Ordovician thrombolites, western Newfoundland'). Since then, my work at GA has focused on the petroleum geology of the onshore Canning Basin and offshore North-West Shelf, and offshore hydrocarbon seepage surveys. I am continuing work on the North-West Shelf to promote the prospectivity of this region and the release of exploration acreage.

'Margaret and I have three children, two of whom are currently studying at ANU; Alice (our Newfoundlander) obtained a BSc at the ANU in 2005 and is now in the third year of an ANU Medicine degree; Robin is in first-year Engineering. Our other son, Julian, graduated from Wollongong University in Performing Arts in 2007, and works as an actor in Sydney.'

Jim Colwell

Jim Colwell undertook his BSc Honours in 1974 and then completed a part-time external MSc through Flinders University in 1979. In 1975, he joined the BMR, working on a variety of coastal, marine geology and geophysics projects around the Australian continental margin and elsewhere, including in the South-West Pacific (Solomon Islands) and offshore Antarctica. In recent years, he has been heavily involved in Australia's submission to the United Nations to define Australia's 'legal' continental shelf as allowed under the terms of the law of the sea convention.

Plate 11.3 As President of the Peugeot Association of Canberra, Brad Pillans sometimes poses as Napoleon at the car club's 'Battle of Waterloo'

Alumni News

Brad Pillans

Descended from a long line of Scottish coalminers, Brad Pillans (Plate 11.3) clearly had a strong genetic disposition towards geology. Recognising this at an early age, Brad's parents gave him a geology hammer for his thirteenth birthday and, from that point onwards, his career path was set in stone, so to speak. At school he was a close friend of Ken Campbell's eldest son, Rod, and it was no surprise that Brad enrolled at the ANU in the very large first-year intake in 1971. Despite his enthusiasm for geology, Brad was somewhat disappointed to receive only a pass grade in first-year geology (and three other passes in chemistry, physics and applied maths). On expressing his disappointment to Professor Campbell, however, the answer came back 'Your early departures from palaeontology practicals did not go unnoticed'!

Brad completed his honours thesis, 'Surficial geology of the Murrumbidgee–Bredbo interfluve', supervised by Keith Crook, in 1974. This was followed by a PhD thesis (upper Quaternary landscape evolution in South Taranaki, New Zealand) in the Geography Department at the ANU, supervised by John Chappell and completed in 1981. Brad lecturered in geomorphology at the ANU from 1979 to 1982, before taking up a lectureship in geology at Victoria University of Wellington in New Zealand (1983–93). He returned to the Research School of Pacific and Asian Studies at the ANU in 1994, before moving to the Research School of Earth Sciences (RSES) in 1998. Brad is currently Professor of Earth Environment in RSES and Vice-President of the Geological Society of Australia (2008–10).

David M. McKirdy

'I came to the ANU from the BMR on a Commonwealth Public Service Postgraduate Scholarship and spent two years full-time there (1974–75), sharing a room with Roger Marjoribanks, Judy Owen and Lesley Wyborn. My internal supervisor was Eric Conybeare and my project was an evaluation of the source rock and petroleum geochemistry of the Amadeus Basin. Upon returning to the BMR in early 1976, I became the first organic geochemist to win the Best Paper Award at the annual Australian Petroleum Exploration Association Conference—ironically, on a topic that had nothing to do with my doctoral research! I eventually submitted my PhD thesis in November 1977, having by then joined the Fossil Fuels Division of the SA Department of Mines and Energy. After a short period at the University of Bristol in 1981, I spent the next 14 months in Oklahoma with Conoco Incorporated working on the application of geochemical techniques to oil and gas exploration in North America and

Indonesia. Between 1982 and 1987, I was a project consultant in petroleum exploration at the Australian Mineral Development Laboratories, before joining the University of Adelaide's Department of Geology and Geophysics as a Lecturer in Petroleum Geology, rising to Associate Professor and serving for three years as Deputy Head of Department. Since 1988, I have presented an annual short course at the Australian School of Petroleum (formerly the National Centre for Petroleum Geology and Geophysics); and in 1999 I taught as a Guest Professor in the Institute of Geology at Cologne University, Germany.'

David has served on many society and symposium committees. He won the Australian Organic Geochemistry Medal in 1996, has supervised more than 50 honours and higher degree students, and has published more than 100 papers and editorial works. His research has encompassed the fields of organic geochemistry, petroleum geochemistry, basin analysis, Precambrian geology and isotope chemostratigraphy, and Holocene palaeolimnology. He has also written on the history of geological teaching and research at the University of Adelaide, where he is currently a Visiting Research Fellow in the School of Earth and Environmental Sciences.

Simon Beams

Simon Beams is the Managing Director/Principal Geologist of Terra Search Proprietary Limited. 'My interest in rocks and minerals developed before I arrived at the ANU. A defining moment was in 1968, just as I was developing my interest in geology, when, as a Boy Scout at the time, I attended a dinner at which Mike Rickard was guest speaker, presenting a talk on his Patagonian expedition to understand why there was a bend in the Andes.

'The years that I spent at the ANU were some of the most enjoyable in my life. I would think that the late 1960s to early 1970s were probably the golden years of the ANU Department of Geology. I felt that, during my undergraduate years (1971–75), the staff profile was at its prime. All our lecturers were world class and intent on passing on their geological knowledge to their students. Unlike today, then, science was a favoured subject for some of the brightest students; the promotion of the mining boom at the time meant that many of these science students often took Geology as an option. The other advantage we had was being in Canberra, where we had access to geologists from the BMR and RSES. All in all, it was an ideal environment for nurturing an interest in geology. All the field trips live long in my memory. On our first mapping excursion to Taemas to map out a fold in limestones—I still have my map—I remember sitting on an outcrop with Professor Campbell and listening in awe as he deduced how the scene might have looked in the Devonian as a stromatolite-strewn tidal flat. I especially remember the field excursions—the South Coast excursion of Professor Campbell, the Mount Dromedary excursion with Ian Smith, the Berridale Batholith mapping project with Chappell and White, and the third-year mapping project at Bredbo. Just as informative were the Student Geological Society trips I went on—in first year to Broken Hill and mapping the Blinman Diapir in the Flinders Ranges with many of the staff; second year to the Warrumbungles and New England; and third year to Tasmania.

'My geological interest got me to university but my overall understanding of geological processes and the skills required to interpret them were well and truly enhanced by my time at the ANU and also by my sojourn at La Trobe University, in Melbourne, which Whitey had instituted as a fairly ANU-dominated 'hard-rock chapter'.

'I had a great time at the ANU; it was also were I met Ellen; we were married a year or so after graduation—a mere 32 years ago. Our married life together has been happy and varied, with three wonderful children and two grandchildren. I know I could not have accomplished as much in my career without the long-suffering support of a close and loyal family.

'I graduated in 1975 with First-Class Honours, which might have surprised a few people who knew me from the Union billiards' room, where I spent a lot of my time in first year! I attribute my final result to a lot of hard work, a great bunch of fellow students who always maintained a high standard, a fantastic field area, full of major undiscovered geological features, and some excellent supervision (Tony Eggleton, Bruce Chappell, Ian Smith, John MacDonald), and technical support. Also, I could not have undertaken my fieldwork without my father, who was more than happy to wander with me around the pretty rugged bush of the hillbilly country inland of the far South Coast of New South Wales.

'In 1976, I worked as an exploration geologist looking for uranium in north Queensland for French uranium explorer Afmeco. The highlight was being camped out for five months straight, in a three-man camp with only drillers for extra company! I was mapping sediments and taking dips and strikes, which would have pleased the non-hard rockers at the ANU.

'From 1977 to 1980, I studied for a PhD at La Trobe University, with Professor Alan White as supervisor. My field area was the 300-km-long Bega Batholith. Most chemical analyses were completed with Bruce Chappell at the ANU. The Bega granites display some of the most beautiful chemical trends of all the Lachlan granites. I was also fortunate and entrepreneurial enough to work with

ANU PhD student Don Hough, who collaborated to organise several contract mapping exercises for the NSW and Victorian forestry departments. One of our studies was on eucalyptus dieback, which we showed to have a largely geological explanation related to heavy clay and poor drainage. After this work with Don, I started paying more attention to surficial geology and processes.

'From 1980 to 1983, I worked with Esso Minerals, notionally based in Sydney, but spent a large percentage of time exploring for volcanic-hosted massive sulfides in the Bredbo–Cooma district. I also spent time with John Walshe at the ANU, who provided some excellent insights into fluid chemistry and ore–mineral relations. As is often the case in exploration, our drill programs encountered only sub-economic base-metal and gold mineralisation.

'From 1983 to 1987, I was based in Townsville with Esso Minerals, primarily involved in exploration for volcanic-hosted massive sulfides, but also exploring for intrusive-related epithermal, mesothermal and breccia gold systems across north Queensland. At the end of my time at Esso, all other geologists had been sacked and I found myself running their entire north Queensland gold and base-metal exploration program.

'In 1987, I formed Terra Search as Principal Geologist and Managing Director, based in Townsville (Plate 11.4). We hit the ground running as a multi-client mineral-exploration consulting/contracting group at the height of the 1980s gold boom. Pretty soon we had a long list of clients, working all over Australia and offshore. Highlights were the discovery of the Reward copper–gold massive sulfide pipe, south of Charters Towers, initially discovered by Terra Search in 1987 working for City Resources. Reward led to the discovery of several other adjacent ore bodies (not all by Terra Search). At recent prices, the new wealth generated by these discoveries is in the order of A\$1.7 billion. Other mineralisation discovered by Terra Search includes the Grevillea Pb-Zn-Ag deposit, south of Century Mine in north-western Queensland; the Gettysberg gold-vein system in the Drummond Basin; the extension of the Mount Mackenzie high-sulfidation Au system, central Queensland; and discovery and/or evaluation of numerous mesothermal and breccia-gold systems in the Charters Towers district.

'Since 1993, Terra Search has developed its expertise in the conversion of hardcopy open-file mineral exploration data into GIS-accessible digital data sets. The mineral-province scale data-compilation projects began with the well-received industry-sponsored Australian Mineral Industry Research Association (AMIRA) projects over the Mount Isa block and eastern Queensland in the period 1993–96. Concurrently, Terra Search began working with the NSW,

Alumni News

Victorian and NT governments, from whom we won competitive tenders to complete compilations in these states. Terra Search has since worked with the SA and Queensland governments as well as Geoscience Australia (GA) in similar compilations.

'In 2002, Terra Search won a 3.5-year, US$1.8 million World Bank project to compile exploration data across the whole of Papua New Guinea, and set up a PNG operation with local geological staff to carry out work at the Department of Mining, Port Moresby. Currently, we are working on a four-year, A$1.8 million exploration-data project across the mineral provinces of Queensland.

'In 1995, Terra Search opened an office in Perth, managed by Director, Dave Jenkins. Our size expands and contracts with the commodity cycle. Generally, there are at least 45–50 geological, field technical and computing employees and management.

'I still derive most satisfaction from field geology and prospect evaluation. I try to squeeze as much interesting geological work as I can in between the demands of managing a large and diverse mineral-exploration service group/ geological consultancy. I have had many satisfying geology jobs in the past 15 years. Most prominent amongst them would be the current work I am doing on the Rocklands project for Cudeco, near Cloncurry, Queensland, which is one of the most significant copper discoveries in the recent boom. Exemplifying the workloads thrust upon us ageing baby-boomers by the shortage of skilled geologists, I personally am producing all the geological interpretations and reviewing well more than 100 000 m of drilling on this project, plus reviewing all the ASX announcements. I am looking over my shoulder to hand over the baton but there do not seem to be many takers! Interestingly, I have employed and worked with a lot of geologists over the past 30-odd years but the two I have ended up working the closest with are both ANU Geology graduates and my fellow Terra Search Directors: Richard Lesh and Dave Jenkins (Plate 11.4).'

Plate 11.4 Richard Lesh, Simon Beams and Dave Jenkins, Directors of Terra Search Proprietary Limited

Doug Mason

Doug Mason completed his PhD on the petrology of intrusive rocks and associated porphyry-type copper deposits of the South-West Pacific over the period 1972–76, with significant assistance from supervisor, John McDonald, the XRF lab of Bruce Chappell and last-minute editing by Jo Heaslip, who was Research Assistant for Ken Campbell and Professor David Brown. Doug and Jo departed together for warmer climes on the north coast of New South Wales, where they built a small house, married, and then Doug took up a postdoctoral position at the University of Toronto, thanks to notification from fellow graduate Ian Smith that the position was available. Doug subsequently took a teaching position for two years at the University of California, Riverside, where he taught mineralogy, optical mineralogy, igneous petrology and ore-deposits geology, and researched magma mingling in the southern California Batholith. They returned to Australia to have a family, and Doug taught mineralogy, optical mineralogy and igneous petrology at the University of Newcastle during 1980–86. Since then, Doug and Jo have built their commercial petrology practice based in Adelaide, providing petrological services to the minerals exploration and mining industries worldwide. Over the past 10 years, Doug's interests have centred on the petrology and thermodynamic modelling of orogenic gold deposits, and Jo has completed a Graduate Diploma in Environmental Science and developed a consultancy in scientific editing.

Mark Stevens

Mark Stevens graduated with Honours in **1976**, then, after 10 years in Sydney (1977–79 with Kratos Minerals NL and 1980–86 with Offshore Oil/Petroz NL), he moved out west with Petroz NL Perth (1987–91). From 1991 to 1994, he worked as senior geologist with Simon Petroleum Technology and then with the Geological Survey of Western Australia, and, from 2006, with the Petroleum and Royalties Division as Senior Field Development Adviser overviewing all petroleum production licences, retention-lease applications and renewals in Western Australia.

Ian Smith

'Having worked as a Demonstrator for several years, I submitted my PhD in **1976** and left Canberra with Lydia, Nic and Meg in the middle of July bound for Canada, where I had been offered a Postdoctoral Fellowship in the Department of Geology at the University of Toronto. Toronto was the largest city in Canada, with a population of three million and very different from Canberra. We rented a house in a rundown neighbourhood close to downtown Toronto where even on a postdoctoral stipend we were relatively well off. Toronto turned out to be a great place for kids and an exciting place to live. My research fellowship was to work on the Archean volcanic-rock sequences of the Canadian Shield north of Toronto. Fieldwork meant that the family spent summers in the beautiful countryside and the opportunity to do research in a completely different environment—a very different prospect from the young volcanoes of Papua New Guinea that I studied for my PhD.

'After two years as a Research Fellow and another as a lecturer at Toronto, I was offered a lectureship in the Department of Geology at the University of Auckland. The attractions of returning to a more active geological environment, and for Lydia and I of returning to New Zealand, outweighed those of North America and we moved to Auckland in September 1979. The University of Auckland has proved a good place to work. There have been many research opportunities in the variety of different volcanoes in the northern North Island as well as projects in the Kermadec Islands, Tonga, Samoa and Vanuatu.

'Lydia completed an MA at the University of Auckland and has had a variety of jobs before settling in at the Auckland University of Technology as lecturer in communication and harassment-prevention coordinator. Nic studied engineering at Auckland University and is now a Don at Oxford University. Meg is a charge nurse at Auckland children's hospital. We look back on our time in Canberra and at the ANU as a particularly happy one.'

W. J. Collins

W. J. Collins was another of the Chappell honours students (**1977**). He left to undertake a PhD at La Trobe University with Professor Allan White, working on granites of the Pilbara region in Western Australia and graduating in 1983. He spent two years in diamond exploration with CRAE before returning to academia as a Postdoctoral Fellow with Professor Ron Vernon at Macquarie University. He took a lecturing position at the University of Newcastle in 1988 and remained there until 2005, when he became a Professor at James Cook University in Townsville.

Bill has studied a range of topics, including granite petrogenesis, Archean tectonics, crustal evolution and geodynamics. His honours work and subsequent publication led to the recognition of A-type granites around the world. He also has published alternative interpretations from Chappell and White on S-type and I-type granite genesis in the Lachlan Fold Belt. This helped revive conventional subduction models for the tectonic development of the fold belt. His recent work on the relationship between geochemistry and geodynamics suggests that most of the Lachlan granites formed in a back-arc setting. This was influenced by work with Ron Vernon on the causes and tectonic setting of low-pressure, high-temperature metamorphic terrains, which are a feature of the Lachlan Fold Belt. With Bob Weibe, he has shown that many plutons are cumulate mushes that build incrementally, much like deposits in sedimentary basins. This has major implications for 'non-constrained' geochemical modelling of granitic systems.

Other highlights include work with Christian Teyssier (Monash and Lausanne universities) on the recognition of the intense and widespread effects of the Paleozoic Alice Springs orogeny in central Australia—contrary to popular belief at the time. He is presently focused on understanding the causes of the break-up of the supercontinent Pangaea. In 2010, Bill was awarded the Carey Medal by the GSA.

Peter Ward

Peter Ward (now Sorjonen-Ward) took his Honours in **1979**. Then, being weaned off graptolites, he worked as a Research Assistant in structural geology with MJR and KAWC. He was informally awarded the ignominious distinction of 'maximum points for minimum words' in exams and essays. In 1981, he went to the University of Glasgow to complete a PhD on the Precambrian rocks of Finland. He claims not to have a career path—'just an interrupted and convoluted trail of uncompleted projects and frustrated ambitions'. Peter has lived progressively backwards through time, from initial studies in the Palaeozoic Lachlan Fold Belt through the Proterozoic of the Outokumpu region in Finland, to the Archean of Fennoscandia and Western Australia. He has been fortunate to work with several research groups in the Geological Survey of Finland and the CSIRO Division of Exploration and Mining in Perth. Throughout this time, his interests have evolved from structural analysis of basin and orogenic processes to mapping and modelling different types of mineral systems, with a particular emphasis on structural controls of hydrothermal processes.

Peter is now based at the Geological Survey of Finland in Kuopio, where he walks across the Archean–Proterozoic boundary twice a day! He recently (2008) organised for John Long (ex-ANU Research Fellow) to lecture on the Gogo fossil fish to the University of Helsinki.

Sue Coote/Slater

'After leaving ANU in **1979**, I spent some time at the University of Queensland before starting work with the Geological Survey of Queensland (GSQ) in 1982. I was initially in the Petroleum Section there, and subsequently joined the Stratigraphic Drilling section, which had the responsibility of drilling deep stratigraphic bores to encourage and aid petroleum exploration in Queensland. Subsequently, this section was renamed Basin Studies. During this time, I completed a Graduate Diploma in Business Administration at what was then the Queensland Institute of Technology.

'I left the GSQ in 1988 to have our first child, Matthew; and my husband, Ray, and I later moved to Collinsville in north Queensland when Ray got a job as mine geologist there. We spent more than 11 years in Collinsville, and our second son, Thomas, was born there, in 1991. During our time there, I essentially did not work in the industry, but was very involved in community and sporting organisations and basically had a great time.

'We returned to Brisbane in 2002, when Ray got a job back in the then Department of Natural Resources and Mines. A few months after we returned, I was asked if I would help out in the Petroleum Section again as they were short staffed. I spent 2002–04 in the Department of Natural Resources and Mines doing technical assessments of petroleum-tenure applications, drilling applications, and so on, before being offered a position with Tipperary Oil & Gas (Australia), which operated the Fairview Coal-Seam Gas Field in central Queensland. I looked after their tenure, land and environmental compliance issues until they were taken over by Santos. In 2006, I started work at Resource Land Management Services, a Brisbane-based consultancy to the oil and gas industries, where I look after tenure issues for several small to medium exploration and production companies, pipeline development projects and various other related issues. Most of our involvement is with coal-seam gas since that is the big-ticket item in Queensland, but I have also been involved in the Zero Gen geosequestration project. Currently, I am a member of PESA and QUPEX and have been on the Queensland/NT PESA committee since 2007.'

Ray Slater

Ray Slater commenced work in **1980** with the Geological Survey of Queensland and spent the next 10 years or so in a variety of roles—mainly engaged on coal exploration—in Australia, New Zealand, Malaysia and Indonesia. He then completed a Graduate Diploma in Mineral Economics externally through Macquarie University in the mid-1980s. In 1990, he joined MIM as Mine Geologist at their Collinsville mine and stayed for 11 years in a variety of roles across all facets of geology (surface and underground), mining, resource/reserve estimation, coal sales, shipping and marketing, and contracts administration.

In 2002, he returned to Brisbane with two grown kids, and briefly rejoined the Queensland Government in the Bureau of Mining and Petroleum, before resigning to start his own business in January 2006—Ray Slater and Associates Proprietary Limited—specialising in tenement management, administration and statutory compliance, and exploration management servicing a range of mainly coal (also minerals and coal-seam gas) exploration and mining clients. The group employs seven geologists and two student geologists (part-time/casual).

Ian J. Ferguson

Ian J. Ferguson, after finishing his BSc (Hons) Degree in the Geology Department in **1981**, worked for a short time with the BMR, including nine months on Macquarie Island. He then completed a PhD at RSES on Marine Magnetotellurics

(MT), supervised by Ted Lilley, and a postdoctorate at the University of Toronto. Since 1990, he has been a faculty member at the University of Manitoba, Winnipeg, Canada, where he teaches a range of courses in the geophysics program and does research in crustal-scale magnetotellurics. He started a term as Head of Department in July 2009.

'I have been fortunate to get back to Australia for a visit most years, partly because my wife, Nancy, is a sedimentologist, and has been working with colleagues from the University of Western Australia on the Canning Basin. We are presently on research leave at the University of Western Australia. I am still working mainly on lithospheric-scale magnetotelluric studies from various Precambrian regions in Canada; however, I have just finished a collaborative MT survey crossing the southern margin of the Yilgarn Craton here in Western Australia—my first land MT fieldwork in Australia since my honours thesis project in the Tumut Trough.'

Mukul Bhatia

'After completing my PhD at ANU in **1981**, I joined the petroleum industry and worked as an exploration and development geologist with Elf Aquitaine in Sydney, and participated in a number of drilling ventures both offshore and onshore. I also took part in the liquified natural gas (LNG) negotiations with Japan, Taiwan and Korea. Then came the 1986 crash of oil prices and I was fortunate to work with AGL in Sydney as a resource analyst assisting the company in a number of financial evaluations leading to merger activities. In 1989, I joined BHP Petroleum in Melbourne and worked on the Timor Sea exploration and development. In 1992, I was promoted to the position of Production Coordinator for Bass Strait and the North-West Shelf Project. This was followed by an assignment to write the development plan of the Bayu-Undan field in the Timor Gap. In 1999, I was asked to visit Houston for three weeks to help the deep-water team. After eight very fruitful years, I am still in Houston, working as a Subsurface Manager for the deep-water developments. In my current job, I am responsible for reservoir engineering and production technology, in addition to the geoscience functions. I thoroughly enjoyed my stay at the ANU. The staff were most helpful and I established friendships with a number of colleagues. The academic standard and expectations were very high.'

Joyce Temperly/Edmonds

'A quick history of me since leaving the ANU in **1981**: I had not even finished my last semester before I had secured a job with CRA Exploration in their

Diamond Indicator Mineral Laboratory in Perth—the perfect job for someone with children to care for. It was a hard decision to actually accept the job because it meant taking my two daughters away from their extended family. We have, however, prospered here in Perth. Both my daughters are now married and have three children apiece, so I am well and truly a grandmother (shudder). You do not get old, but your kids sure do.

'I continued working for CRA until it became Rio Tinto, but shortly after having seconded me to Thunder Bay in Canada for a month, they discovered that they did not need me anymore. After that, I continued to work in the diamond exploration industry in a laboratory capacity for another nine years until diamonds went into decline in mid-2007. A year with nothing; then I secured a place with the Geological Survey of Western Australia in their Statutory Exploration Information Group, where I continue to work.

'I have been heavily involved with the folk music and the dancing scenes since my arrival in Perth and that has enabled me to see many different parts of the country—and I even made a trip to Hong Kong, performing with various dance groups; and I have danced at the National Folk Festival five times—twice since it has been in Canberra (the last time was in 2007). I have also added another degree to my list of achievements. After being retrenched by Rio Tinto, I went to Curtin University of Technology and completed a Bachelor of Arts (Hons) whilst at the same time dashing backwards and forwards to work for Ashton Mining in their diamond indicator laboratory. Phew!

'My only interesting geological achievement was the discovery of a new mineral—"Arsenoflorencite-Ce: a new arsenate mineral from Australia" (E. H. Nickel, J. E. Temperly, *Min. Mag.*, October 1987). I was trying to decide whether it was a zircon or not when I noticed that the crystal structure was trigonal, whereas zircon is tetragonal. So off to Ernie Nickel at CSIRO to check it out after our SEM failed to give us a definitive answer. As for me, I am now Joyce Edmonds because I remarried early last year.'

Paul Habelko

Paul Habelko is Manager of Strategy and Planning for Chevron Global Gas. Paul graduated from the ANU in **1981** with a BSc Degree (Hons) in Geology. He also holds an MBA from Macquarie University. Upon graduation, Paul joined Exxon as a geophysicist; then, after completing another MBA at INSEAD Fontainebleau, France, in 1991, he worked with Gaffney, Cline and Associates based in London, as senior consultant in the international energy arena, and was engaged on numerous projects around the world, including several for the World Bank and major financial institutions.

In 1993, he joined Statoil in Stavanger, Norway, and held positions with the International Exploration and Production Company including Planning Manager, International New Ventures Manager and Commercial Manager for the company's Russian activities. Paul joined Chevron Corporation in 2002, initially as the Commercial Manager for the North-West Shelf Project based in Perth, Australia. He took up his current position as Manager of Strategy and Planning at Chevron's US Corporate Headquarters in May 2005. In this capacity, he oversees planning and strategy for Chevron Global Gas and its six Strategic Business units, which include the corporation's Liquid Natural Gas and Gas to Liquid, shipping, power generation, pipeline and gas-trading businesses. Paul is a longstanding member of the Association of International Petroleum Negotiators (AIPN) and the American Association of Petroleum Geologists (AAPG). He was also on the board of the Australian–American Chamber of Commerce in San Francisco.

Although they 'still call Australia home', Paul and his wife, Joanne, and two sons, Freddy and Theo, currently reside in Danville, in the Bay area of San Francisco, California.

Gordon Taylor

After completing his Honours Degree in **1982**, Gordon Taylor worked with Keith Crook as a Research Assistant. He was then recruited by the BMR to work on the Regolith–Terrain Map of Australia. Gordon then left geology to pursue his interests in filmmaking, and completed a three-year degree at the Australian Film and Television School in Sydney. His work has been shown on SBS and BBC television. Subsequently, he was recruited by the Science Unit at ABC Radio and began a 20-year career in journalism.

At the ABC, Gordon has worked as a reporter in radio and television. He spent two stints as the ABC's Science and Medicine Reporter, and three years living in India as the ABC's South Asia Correspondent. He has managed the ABC Radio current affairs program PM, and was Manager of the Radio National network for three years. He currently lives in Canberra and works in the Federal Parliamentary Press Gallery for the ABC, as well as producing stories for the *Stateline* and *7.30 Report* programs.

Geology at ANU (1959–2009)

Doone Wyborn

Doone Wyborn worked at BMR/AGSO in the 1970s mapping the Tantangara, Brindabella and Araluen 1:100 000 sheets with Mike Owen. Granite mapping on the adjacent Berridale sheet was being done concurrently by ANU students, and the whole I-type and S-type story was being unfolded. He studied at the ANU from 1979 to **1983** for his PhD on 'Fractionation processes in the Boggy Plain zoned pluton'. This work was extended to make important distinctions between the Silurian (cordierite bearing S-types) and Devonian volcanics in the Canberra–Yass region. From 1997 to 2000, he worked as a Visiting Research Fellow at the ANU, partially supported by AMIRA and a grant from the Australian Greenhouse Office to drill a 2 km well in the Hunter Valley searching for hot rocks. This work led to commercial opportunities (see the section on hot-fractured rock, Chapter 6). Doone is now Executive Director, Science and Exploration, for Geodynamics Limited.

Michael Andrew

'After the mutual relief between myself and ANU staff on finally completing my degree in **1983**, I was about to embark on a culinary career; however, my mother spotted an advertisement for a position with Coffey & Partners, and, despite the lack of a suit, I was able to get the job. I spent the next four years climbing in and out of the test trenches, footings and foundations of civil engineering projects in and around Canberra, plus a four-month stint at Ok Tedi in Papua New Guinea drilling karstic limestones for a new tailings-dam site; surprisingly, the tailings dam was not constructed. A chance encounter at an airport bar led to a job exploring for gold in Queensland, New South Wales and Victoria for the next 18 months. A call from an old colleague saw me moving to Whyalla, South Australia, and working for BHP Steel for the next five years on iron-ore projects in the Middleback Ranges, where, against my better judgment, I got involved with computers, having been able to avoid them to date. I then joined Normandy Mining based in Adelaide, getting involved in resource estimation and evaluation, managing to avoid retrenchment, riding the boom–bust cycles of the resource industry. Newmont acquired Normandy and, after 10 years, I was lured to the dark side and joined Snowden as a Principal Consultant, based in Perth but getting to see a lot of the world. I recently completed a Postgraduate Certificate in Geostatistics, finding the excuses I used 20 years previously at the ANU still worked! During this time, my most rewarding project has been meeting my wife, Louise, and the arrival of our daughter, Lily.

Bruce Turner

Bruce Turner graduated with First-Class Honours in Geology in **1983**. The year after, he worked for a mining company in Canada and then at the BMR back in Canberra, where he wrote up and published his honours work in collaboration with Doon Wyborn and Bruce Chappell. After considering the possibility of pursuing a PhD in Geology, he decided instead to do a Dip. Ed. at the University of Melbourne in 1985. He then taught maths and science (including VCE physics) for two years at a state high school, where he found the camps and co-directing the school musical particularly rewarding!

Then, in 1988, he entered a one-year, 'fast-track' scheme for graduates with the Victorian Public Service and ended up working as a Policy Adviser in resources policy with the Department of Premier and Cabinet. He also met his wife, Lindy, in the same scheme; they now have two teenage boys. Bruce worked for the Victorian Government for 10 years, mainly on environmental-impact assessments and reform of the State's land-use planning system. In the course of this work, he discovered an interest in mediation and, in 1998, he was awarded a Churchill Fellowship to study 'public-dispute resolution in resource use and environmental issues' in North America and the United Kingdom. After the fellowship, he tried his hand working as a facilitator and mediator in the private sector. For the past seven years, he has been working independently from his Melbourne home as a freelance facilitator for a wide range of government and private-sector clients, running public consultation and consensus-building processes.

He continues to find his background in geology very useful in his role as a bridge between technical experts and the general public in matters involving complex, and often uncertain, science.

Jane Rodgers

'After graduating from the ANU in 1983, I moved to Sydney to work for Esso Australia in their Petroleum Exploration Department. Initially, I worked on the Galilee Basin, mapping oil and gas prospects, but then moved to the Browse Basin as a seismic interpreter. I then became involved in a multidisciplinary team conducting a regional study of the petroleum systems of the Cooper–Eromanga basins. In 1989, I moved to Houston, Texas, and participated in a regional study of the Paleozoic basins of Africa and South America. Later, I moved to Norway and was involved in the development of the Snorre, Vigdis and Statfjord fields in the Norwegian North Sea.

'In 1993, I relocated to Melbourne and commenced work in the Gippsland Basin conducting three-dimensional seismic interpretation over the Bream and Tuna fields and providing input to field development plans. I moved to the Delhi Petroleum Cooper–Eromanga Basin group in 1997 and monitored the exploration and development program of the operator, Santos. More recently, I have been working on exploration permits in the Gippsland Basin with a focus on deep-water targets.'

Jon Olley

Jon Olley gained his BSc in Geology (Hons) at the ANU in **1984** and joined the CSIRO as a technical assistant in the Baas Becking Laboratory in Canberra. From 1990 to 1994, he completed a CSIRO-sponsored PhD at the University of New South Wales, on the use of uranium and thorium decay-series radionuclides in sediment tracing. He subsequently held various positions, ranging from Research Scientist to Research Director, conducting and directing research focused on understanding how large catchments respond to changes in land use and climate. He has authored 86 scientific papers, co-authoring three papers in Nature and two in Science. In March 2007, Jon became Deputy Chief of the CSIRO Division of Land and Water. The division has about 500 staff and visitors and an annual budget of about $70 million. During his time with CSIRO, he has maintained his university connections and has supervised seven successful PhD and six honours students, and is currently supervising four PhD students in their final year. Since 2005, he has also been an Adjunct Professor at Griffith University in Queensland. In August 2008, he left the CSIRO to take up a new and tenured appointment as Professor of Water Science in the Australian Rivers Institute at Griffith University.

Jon married a fellow ANU geology student, Vanessa Grey, and they have two boys, now twenty and twenty-two years old. Vanessa works as a special-needs teacher in preschool, helping integrate special-needs children into the school system. She delights in enthusing the preschoolers with a love of exploring how the world works. Her collection of rocks and fossils gets a regular workout.

Robert A. Creaser

Robert A. Creaser graduated in **1984** and, after a term position at the BMR with Lynton Jacques, he moved to La Trobe University to undertake his PhD with Allan White. His main lines of interest were in geochemistry of the solid Earth, and, during his PhD studies, he became interested primarily in radiogenic-isotope geochemistry, studying with C. M. Gray at La Trobe, and J. A. Cooper in Adelaide.

In 1990, he moved to Los Angeles for a postdoctoral position at the Division of Geological and Planetary Sciences, to work with Professor G. J. Wasserburg in the 'Lunatic Asylum' of the Charles Arms Laboratory. Here he developed new methods for the isotopic analysis of the platinum-group elements—a project that would shape his future research career. In 1992, he accepted an Assistant Professor position at the University of Alberta, Edmonton, and was promoted to Associate Professor in 1997 and Professor in 2000; he served as Associate Chair (Research) in the Department of Earth and Atmospheric Science from 2003 to 2008. His main area of research involves the application of the rhenium-osmium (Re-Os) isotope system to crustal geochronology, and he has developed methodologies to provide precise and accurate dating for crustal materials such as shales, sulfide minerals and oil. His research has been awarded the Geological Association of Canada's W. W. Hutchison Medal and the University of Alberta's Faculty of Science Research Award. He is a Fellow of the Society of Economic Geologists, was awarded Distinguished Fellow of the Geological Association of Canada in 2007, and was elected as a Fellow of the Royal Society of Canada in 2008. Outside his academic life, he enjoys golf, skiing and gardening and has achieved Shodan (first-degree black-belt) status in jujitsu. Rob and his wife, Pauline, have a daughter, Emily, and a son, Will.

Oliver Raymond

Oliver Raymond worked on the Mount Wright gold project in north Queensland for his Honours Degree in **1985**. He was then a geologist with MIM at Mount Isa until 1988. While there, he escorted a GSA tectonics group excursion underground to view the structure and mineralisation. From 1989 to 1990, he took an MSc from the University of Tasmania working on the Mount Lyell deposit. Since 1991, he has been working with AGSO/GA on a variety of projects including the Mount Isa Inlier, the Lachlan Fold Belt, Gawler the Craton, national geological maps and, currently, on international data modelling. Ollie has fond memories of his time at the ANU, especially fieldwork and petrology, and his companions in the honours lab: 'six walking zombies' by submission time. He edited the 1984 student geology magazine, *Lithenea* (see Plate 8.3).

Angela Hume/Thorn

'I graduated in **1985** when I was fifty-two. Jobs in geology were not really on offer except on the North-West Shelf. Plenty of people with PhDs were on the dole, which meant I had to fall back on maths and computing. So I celebrated my freedom to become employed full-time and continue to live with my family by taking a 'safe' job in the Public Service. Having survived the 1987 purge and deconstruction, I eventually wrote and managed a large computer database—this in the days before computers had mice! Grown-up children and actual money meant more freedom to travel with my husband in Australia and overseas. Then, and now, I have regarded myself as an amateur geologist. This has dictated destinations for travel (for example, Broken Hill, north-western Scotland) and picnic spots (any quarry anywhere). I was, and still am, attracted by the beauty of rocks and crystals and the landscapes they inhabit. Now in retirement, I still enjoy travelling and accumulating a geological library.'

Paul Johnston

'After finishing my PhD at the ANU, I joined the Royal Tyrell Museum of Palaeontology in **1985** as Curator of Invertebrates. My main duties were to establish the exhibits on Precambrian and Palaeozoic life, and to build an invertebrate collection. [His walk-on display of early Burgess shale faunas is brilliant—MJR.] In 2004, I left the museum to take a faculty position in the Department of Earth Sciences at Mount Royal College in Calgary—Canada's newest undergraduate university. In terms of research, I spend most of my time studying bivalve evolution, but recently I have developed a new hypothesis that the Burgess shale communities were mostly chemosynthetic organisms focused at sea-floor brine seeps.'

Jyrki Pienmunne

'After graduating halfway through 1985, I left Australia and managed to get a job with the Geological Survey of Finland as a Research Assistant. I was based in the Southern Finland Office at Espoo and worked in the Exploration Department, exploring for gold and base metals.

'In 1989, I returned to Sydney, where I completed a Graduate Diploma in Minerals Exploration part-time at Macquarie University, whilst working in the construction industry. In 1992, I went back to Finland to once again look for base metals and gold in southern Finland. Then, in 1994, I returned to Sydney yet again and, after a few months in the construction industry, got a job with the Geological Survey of New South Wales, where I spent most of my time in the Land Use and Resource Assessment Unit dealing with matters requiring geological input, and doing commodity studies in industrial minerals and construction materials.

'Concurrently, I completed an MSc part-time at Macquarie University by 1997. My thesis was 'On the geochemistry of the Tomago Aquifer'. My supervisor at the time, Professor Blair Hostetler, convinced me that doing a PhD would be good idea and this involved a field trip to Cobalt, Ontario, in 1998. I was supposed to study the trace-element and stable-isotope geochemistry of the Ag-Ni-Co ores that occur at Cobalt with the aid of laser-ablation ICP-MS and multi-collector ICP-MS equipment, with the purpose of shedding light on where the ore-forming fluids had come from. It was a good project but due to family and work commitments it ground to halt in about 2003. On the upside, I did enough experimental work on a variety of antimony mineral samples from around the world to present a poster on the natural variation of antimony isotopes at the Applied Isotope Geochemistry Conference held in 2001.

'In 2004, the NSW Government decided to relocate the Department of Mineral Resources, and the Geological Survey with it, to Maitland in the Hunter Valley. I elected not to move and was made redundant with a host of other similarly minded people. In early 2005, I commenced working for Triako Resources in North Sydney. Triako, at the time, operated a copper-gold mine at Mineral Hill north of Condobolin and was developing the Hera gold-base-metal deposit near Nymagee. Triako was taken over by CBH Resources in late 2006, and, as a result, I joined them. CBH is an Ag-Pb-Zn miner and mine developer with interests in the Cobar Basin, Broken Hill and Western Australia. Due to the economic downturn, CBH reduced its exploration staff (including me) in mid-2008 and, since then, I have been contracting for the highest bidder (or any bidder, as the case may be) and am currently working for CBH again!'

Tim Munson (Plate 2.43)

'While at the ANU, I completed my PhD from the University of Queensland in 1986. After 15 years as Tutor and Museum Curator at the then Department of Geology, I left to join the NT Geological Survey in early 2001. I was the Editorial Geologist from 2001 until 2005; since then, I have been in my current role as Project Manager, Publications and Graphic Design. In that time, I have edited and supervised the production of some 180 geoscience publications, including the substantial *Timor Sea and Central Australian Basins Symposia Proceedings* volumes (combined total of 56 peer-reviewed papers), plus numerous commodity volumes, geological maps and explanatory notes. I was lead Editor of the volume on the *Geology and Mineral Resources of the Northern Territory*, due for release in March 2009. This is the first time that the geology of the territory had been summarised in a single volume.

'In 2009, I wound up my current editorial duties after our annual conference in Alice Springs (AGES in March) for a new role involving sedimentary-basin analysis. I was really looking forward to leaving the service/support roles (for example, museum curator, editor) behind me and making a more direct geological contribution. Darwin is a nice place to live and work in; the city is like a very large country town, but with capital-city facilities. We live in the satellite city of Palmerston, about 30 km from the Darwin CBD. The Geological Survey is a great place to work; it is well funded, has excellent staff and facilities, and produces some high-quality geoscience. I feel at home here and have flourished. The ANU was still a most enjoyable and unforgettable experience.'

Simon Veitch

'After completing a BSc at Monash University with majors in Zoology and Geology (1977–80) and vacation employment as a Seasonal Ranger for the then Victorian National Parks Service, I worked as a coal-exploration geologist for BP Australia Proprietary Limited (1981–83) assessing prospects in central and northern Queensland and in central New South Wales. Then, after travelling overseas in 1984, touring Africa, Europe and South-East Asia, I completed a Graduate Diploma (Geology and Geomorphology) at the ANU (1985–**86**) with part-time work tutoring and as a Research Assistant to Keith Crook. From 1986 to 1990, I worked as an exploration geologist with RGC Exploration Proprietary Limited on gold and base-metal prospects in central New South Wales, including development of the Lucky Draw goldmine at Burraga, New South Wales. In 1990, I joined what is now the Australian Government Department of Agriculture, Fisheries and Forestry in the National Resource Information Centre (NRIC) to develop geographic information systems and convert Australia's hardcopy natural-resource maps to digital data sets (including soils, geology, regolith, hydrogeology, geomorphology and bathymetry). My other roles in the department have included managing a project to identify a site for an Australian national low-level radioactive waste repository (1992–95)—work that was recognised by the then Australian Mineral and Energy Environment Foundation with a Team Award in 1994. From 1996 to 2003, I worked in the department's Bureau of Rural Sciences, leading spatial-information, decision-support and land-management programs. I graduated from the Australian Rural Leadership Program in 2003. From then until 2006, I managed policies and programs in the department's Natural Resource Management Division, with responsibilities including soils, weeds, feral animals, salinity and government–industry natural-resource management sustainability partnership initiatives.

'In 2006, I joined the Fisheries and Forestry Division as Manager of International Fisheries, with responsibility for Australia's engagement in international fisheries management organisations, including the Commission for the Conservation of Southern Blue-Fin Tuna, the Western and Central Pacific

Fisheries Commission and the Indian Ocean Tuna Commission. With my wife, Liz, and three teenage children, Stuart, Josephine and Lachlan, we live and work on our small sheep and cattle farm at Burra, New South Wales, about a half-hour drive south of Queanbeyan and Canberra—still looking for rain!'

Herman Voorhoeve

'I fondly remember your [MJR] lectures and practicals in structural geology and the field trip we did in the vicinity of Wee Jasper. After graduating in **1986**, I worked for a number of years as an exploration geologist with CRA Exploration, looking for gold in Kalimantan, Sumatra and north-western Queensland, and base metals in the Great Sandy Desert. Eventually, however, I did an MBA and have been involved in downstream sales and marketing with Mobil Oil since then. So, I am afraid, it has been a few years since I have been out into the field with a geo-pick and a hand lens.'

Monica Yeung

Monica Yeung graduated from ANU Geology in **1986**. She started work in 1984 at the BMR in the Palaeogeographic Maps Project. In 1991, she started her own geo-tourism company, Gondwana Dreaming Proprietary Limited, and left the BMR. Since then she has been interpreting geology and its relevance to everything from wine to climate change on Australia-wide and international tours. Since 1993, Monica has also been involved with the Age of Fishes Museum at Canowindra, New South Wales, where thousands of fish fossils have been found at a site where the fish died during a drought 360 my years ago. Monica has been organising fossil digs to raise funds for research and for the museum and in recent years she has been serving as a board member to help the museum.

Kristina Ringwood

Kristina Ringwood completed a Graduate Diploma at the ANU in **1987**, and then began working as a geologist with Western Mining Corporation (WMC) for five years, before obtaining study leave to complete a Masters in Environmental Science at the University of Melbourne, researching arsenic management in a tailings dam. Following that, she worked in corporate environment roles with WMC, including managing the industry's leading public environmental report at the time on 'climate-change policy, and environmental-management systems'. A secondment then followed to the World Business Council for Sustainable Development in Geneva to work with global business leaders and international conservation organisations on environmental policy, including water and biodiversity.

'I am currently developing and implementing a global water strategy for Rio Tinto. The strategy positions the company to manage water risk and opportunity over the longer term. This involves work with mine sites around the world as well as engagement with external stakeholders.'

Steve Sheppard

'After completing my BSc (Hons) in mid-**1987**, I worked as an exploration geologist for a joint venture east of Nullagine in the Pilbara from September 1987 to January 1988. During that time, I decided to return to further study, starting a PhD at the University of Western Australia in March 1988 looking at platinum-group element mineralisation in the Windimurra intrusion in the Yilgarn Craton. This project fell through, so I started another project looking at gold mineralisation, granites and alkaline-igneous rocks in the Mount Bundey area south-east of Darwin. After finishing my PhD in April 1992, I graduated in the midst of an industry downturn. Following a period of unemployment and child rearing (and hitting golf balls into the rough on various public golf courses in Perth), I started work in June 1993 with Placer Pacific as a near-mine exploration geologist at Granny Smith in the north-eastern Goldfields. In January 1994, I started work in the Regional Mapping Branch at the Geological Survey of Western Australia. During the past 14 years, I have been involved in regional mapping and the interpretation of whole-rock geochemistry in the Kimberley and Gascoyne regions.'

Michael Conan-Davies

Michael Conan-Davies worked at Olympic Dam for his honours project, and then, after graduating in **1987**, did a year of exploration in the forests of British Columbia. This was followed by four years of exploration with WMC in the Tanami Desert, at Pine Creek and Eyre Peninsula before another stint at Olympic Dam. 'I then put my geology hammer in the cupboard and took on Business Development and Strategic Consulting.'

Pauline English

Pauline English came to the ANU as a mature-age student with good experience of bush walking. She worked on Mount Bogong in the Victorian Alps for her Honours Degree in **1988**, and then spent a year as a Research Assistant with Bruce Chappell, before taking a PhD (2002) at RSES on 'Cainozoic evolution and hydrogeology of Lake Lewis basin, central Australia'. She then worked for several years with the CSIRO Division of Land and Water, before joining GA in the hydrology section.

John Stanner

John Stanner graduated in **1990**. 'I stayed on in Chile (Argentina and Peru) during the past 17-odd years and have made it my home. I was a field geologist in the 1990s, and then went back to university for my MSc at the Royal School of Mines in London; now I understand the interior decor of the Old Geology Building at the ANU! I am more on the financial end these days (I knew that BEc would eventually come in handy), jetting about the planet looking for ever-scarce resources (at least in places where one feels safe to dig them up). I run a couple of businesses and am setting up an investment fund in London. Of more geological interest, I was a founder and owner of the Boomerang Pub in Santiago for 10 years from the mid-1990s (home to the local Mining Club). The mining world continues to be a great (and small) world to live in, more international than anyone could imagine and more real (that is, away from the five-star hotels) than almost any other job. At one stage, I counted that I had spent the equivalent of three full years in a tent—sometimes in the central plaza of a poor Peruvian village high in the Andes. Hopefully, such stories will inspire a future group of students to take up the subject. I would highly recommend that any economic geology student makes the trip to Chile/Peru at some time in their early careers, as there is no amount of classroom explanation that makes up for seeing a 2000 m profile through an epithermal system into a porphyry.'

Mark Gordon

Mark Gordon transferred from the University of Canberra to take Honours at the ANU in **1990**; then, after a few months as a technical assistant for Dr Ellis, he worked for Geopeko at Parkes. From 1991, he worked in Townsville and Brisbane with MIM Exploration searching for base metals in the Georgetown Inlier and then on a copper project near Molong, New South Wales. In 1995, he joined Great Fitzroy Mines, working in Namibia and Greece, before returning to head office in Sydney. From mid-1999 to late 2001, he did contract work from Canberra, in the Ordovician of New South Wales and the Archean in the Pilbara. In 2001, he left for Oman, where he spent three years drilling out a Cyprus-style VMS deposit for the National Mining Company. Here, he met his wife and was married by the Romanian Ambassador in Abu Dhabi, before returning to Australia in December 2004. After a year in regional exploration for the CSA mine in Cobar, he became chief geologist for Copper Resources Corporation, working mainly in the Philippines. Since 2007, he has been with Taylor Collinson as a resources research analyst in Sydney.

Geoff Deacon

'After graduating in 1990, I completed my PhD on Quaternary micropalaeontology and sedimentation in the Joseph Bonaparte Gulf, Western Australia, in 2000 at the University of Western Australia. I am now living a very happy, widely varying, full life in Western Australia, currently lecturing at Murdoch University but also managing a mineral collection at the WA Museum, as well as researching aspects of mega-fauna, which I have helped collect from Nullarbor cave sites over the past four years. I spent 2006–07 as the Regional Manager of the Geraldton Museum—a position and location that I enjoyed immensely. I am currently on the council of the Royal Society of Western Australia, and, until recently, was the President of the University of Western Australia Geology Alumni, which I helped found in 2002.

'I still get out for the occasional exploration contract, and recently spent two months exploring the wilds of north-western Tasmania for potential VMS-style deposits. I spend as much time travelling in my spare time as possible. Please pass on my regards to all at the department who might remember me.'

Jeremy Peters

'I have managed to combine my ANU Geology (**1991**) with Mining Engineering (WA School of Mines, 1996); incidentally, I lectured first-year geology there with some success, failing only three malcontents in two years, and am currently consulting in Perth. I see some of the others around the traps here occasionally, mostly keeping in contact with Antony Shepherd.

'Despite an unnatural attraction to dark, smelly holes in the ground, I have not sold out to mining completely, as I manage to keep a smaller clientele happy with my modest abilities in both disciplines, finding that "geos" do not always understand the drivers of mining and engineers certainly do not adequately contemplate the geology of the deposits they mine. This suits junior mining companies and institutions, as I find that I can comfortably cover both the engineering and geological aspects of a project. I even have a couple of grassroots exploration projects on the books, which suits me fine, as not only do I get to roll my swag occasionally, I get to see more rocks and, as Dave Ellis once put it: "The more rocks you see, the better you are."

'If I can offer my view of the ANU: although it (was—is?) directed towards preparing students for higher degrees and academia rather than industry (cf. the number of my contemporaries with PhDs), this was its greatest strength, in that I find that my theoretical background and understanding of the chemistry, relationships and mechanics of deposits and the rockmass in general is far better developed than that of contemporaries from other, more practical institutions. You learn the theory at the ANU and pick up the practice on the job! Not a bad thing at all.'

Dennis Charles Franklin

Prior to enrolling at the ANU, Dennis Charles Franklin served in the Royal Australian Navy as a supply and logistics specialist, from 1981 to 1989. He considers the most important roles he played were operational support of the Darwin Patrol-Boat Squadron and capital-project management. He has continued his military career as an Intelligence Analyst in the Navy Reserve and holds the rank of Lieutenant Commander.

In 1991, Dennis completed an Honours Degree in Geology at the ANU, followed by a PhD (1997) at the University of Tasmania. He was awarded a full scholarship to conduct research into complex-systems behaviour in Prydz Bay, Antarctica—work that was supported by the Australian National Antarctic Research Expedition. Dennis studied the influence of climate-driven sedimentological processes on the geological record as expressed on the continental margins of the Antarctic continent. He has published a number of articles and conference papers and is particularly proud of the discovery of a new order of nano-fossils that can be used as a palaeo-environmental indicator of cryogenic deposition.

From 1997 to 1998, he was employed by Grant Geophysical on a program of exploration for commercial natural gas reserves in Bangladesh. This included the management of the logistical support facility in Chittagong, and the management of large seismic survey crews in the field. From 1998 to 2000, he returned part-time to the Navy as a Staff Officer in the Maritime Warfare Training Unit supporting the activities of five warfare schools at *HMAS Cerberus*.

Since 2000, he has held various positions at Computer Sciences Corporation (CSC). The first of these roles was as a knowledge broker supporting the delivery of IT services to BHP Billiton. Next, he took on the role of Senior Consultant on Competitive Intelligence, Data Management, Information Management, Knowledge Management, IT Strategy, IT Governance, Radio-Frequency Identification (RFID) in the copper electro-winning refinery environment, RFID in support of logistics operations, and in support of underground-vehicle traffic management. Next, he held sales and account management roles as Account Executive for BHP Billiton's Australian Mining Operations, and Account

Executive for Corporate Systems, where he managed a business portfolio of more than $50 million. He is currently the Director of CSC's Natural Resources Centre of Excellence, which has made several notable achievements. He also represents CSC's interest as a member of the board of CRC Mining.

In 2006, he completed a Master of Business Administration at the University of Queensland (Mt Eliza Business School), and a Certificate in Executive Leadership from Cornell University in 2007. Dennis has been awarded an Australian Defence Medal, a Defence Long Service Medal, the Lonsdale Medallion (Navy Reserve Staff Course Dux), and CSC CEO Outstanding Achievement Award, 2004.

Cameron (Cam) Schubert

Cameron (Cam) Schubert graduated from the ANU with BSc (Hons) in Geology in 1991. 'My first job out of university was with Carpentaria Exploration Company—part of the Mount Isa Mines Group. I was lucky enough to spend the next 16 years transferring around the MIM and X-Strata operations. I followed my structural geology interest into the field of mine stability and geotechnical and rock-mechanics engineering. I completed a part-time MAppSc (UNSW) in Geological Engineering (1996) and a MEngSc (University of Queensland) in Mining Geomechanics (1998), and worked in these fields at Mount Isa, the McArthur River Mine (Northern Territory) and the Ernest Henry Mine before ending up in management roles. In August 2007, I joined BHP Billiton to work on the Olympic Dam Expansion Project.'

Robert Corkery

Robert Corkery took his BSc (Hons) in 1992, supervised by Professor Ted Lilley of RSES. In 1998, he gained a PhD in Physical and Material Science from the Applied Maths Department (Research School of Physical Sciences) supervised by Professor Stephen Hyde. 'You might say things were not stacked in my favour going through undergraduate years at the ANU (1985–91). Those were somewhat uncertain times for my future. I recall staff had to go against their instinct in not giving a break to "strugglers". This played some part in decisions to see me through to honours and eventually a PhD; I am glad for that.

'A real turning point came for me when Ted Lilley was my mentor during honours and the time was right to take things seriously, and things worked out with a First-Class Honours. Professor Barry Ninham was also a great mentor during my PhD. Incidentally, one of the other turning points for me was taking a year off after second year to work with WMC at Mount Magnet (Hill 50, underground) and then at Paragon Gold. The miners and supervisors were tough blokes and the mining game was a serious one. It taught me how to do an honest day's work. Working in industry for five years showed me the academic world was not all there was to life. But now I have struck a balance at the institute at which I work; we try to bring academic work to industry, and our funding is roughly 60 per cent industrial and 40 per cent academic. Various mining and resource companies are on our membership, and I like working on finding new sources of clays, salts and other minerals for the consumer industry. My main game these days is material science—perhaps as a result of all that hard-rock training. My career moves have been as follows: 1998, Postdoctorate in Physical Chemistry at Lund University, Sweden; 1999–2004, Research Scientist with Procter and Gamble Corporate Research, Cincinnati, USA; 2004–07, Research Scientist at YKI Institute for Surface Chemistry, Stockholm, Sweden; 2007–08, Research Director at YKI.

Chris De-Vitry and Claudia Camarotto

Chris De-Vitry (**1991–92**) studied petrology, structural geology, palaeontology, sedimentology and economic geology at the ANU, and then took Honours in Geology at the University of Western Australia, followed by an MSc. He is currently completing an MSc in Geostatistics at the University of Adelaide. **Claudia Camarotto** gained her Bachelor of Economics and BSc (Petrology, Economic Geology, Geography) in 1988–**93**, followed in 1996–97 by a Diploma in Human Resources. She is currently studying for a Diploma of Accounting.

Chris has worked in several positions: 1993–2001, for WMC Limited (Geological Assistant, Mine Geologist, Ore-Resource Geologist, Senior Ore-Resource Geologist); 2001–03, BHP Billiton at Newman as Senior Ore-Resource Geologist; 2004–06, at Stawell Gold Mines, Victoria, as Senior Ore-Resource Geologist; 2006 – present, Principal Consultant for Quantitative Geoscience. Claudia has worked as a Research Assistant at the ANU (1993); for WMC Limited (1994–2001) as a Business Analyst, Human Resources, (Standard Accounting Practice) and Year 2000 Project teams; 2001 – present, as a Financial Officer with Newman Women's Shelter.

Chris and Claudia met in second-year petrology and were dating by third-year petrology, and they married on 31 December 1994. They now have three children (Nicolas, nine, Anton, six, and Amelia, two). Since marrying, they have moved almost every two years, but are hoping the next move (in October 2008) to an acreage on the outskirts of Brisbane will be permanent. They have lived in Leinster, Three Springs, Newman and Perth, in Western Australia; Melbourne and Stawell, Victoria; and Toronto, Canada.

Chris is still a keen cyclist and kayaker and Claudia's life revolves around kids and holding the fort while Chris is travelling. She has recently developed a carbon emissions conscience, will recycle you if you stand still long enough, and walks everywhere—which is handy given the price of petrol! They try to visit the east at least once every two years to see family in Canberra and New South Wales.

Penny King

Penny King worked on A-type granites with Professor Chappell after graduating with First-Class Honours in **1992**. She then took a PhD at Arizona State University before joining the staff of the University of Western Ontario as Assistant (1999–2006) and later (2006) Associate Professor. She is currently a Senior Research Scientist at the Institute of Meteoritics at the University of New Mexico. In 2005, she was a Distinguished Lecturer for the Mineralogical Society of America.

Antony Shepherd

'I am currently living in Perth and working for Barrick Gold. In fact, with takeovers and mergers of various companies, I am still working for the same company I started with when I graduated from the ANU in **1992**!

'After graduating, I first worked as a Research Assistant for Bruce Chappell, and then joined Geopeko for a few months at Parkes, New South Wales. From 1992 to 1997, I worked at Meekatharra and Mount Morgan for Dominion Mining Limited and Plutonic Operations Limited. In 1998, I became Senior Geologist with Barrick Gold Corporation and Homestake Gold of Australia, still working in Western Australia. From 2002 to 2005, I worked in Canada for Hugh River Gold Mines Limited; then, in 2005, as a consultant in Sichuan, China. From November 2005 to the present, I have been Superintendent Mine Exploration–Australia/Pacific RBU, based in Perth.'

Chris Pigram

Chris Pigram was a PhD student from 1988 to **1993**, after which he worked with BMR/AGSO/GA and is now Deputy CEO and Chief of the Geospatial and Earth Monitoring Division.

Megan Spandler/James

'Having enjoyed my studies at the ANU and graduating with Honours in **1993**, I decided to share my love of science with others, so I became a secondary schoolteacher. I find it challenging and rewarding teaching twelve–eighteen-year-olds in several aspects of science and especially geoience.'

Clinton James Rivers

'I majored in geology and regolith studies, with sub-majors in chemistry and Indonesian, and graduated with Honours in **1993**. I was among one of the early batches of CRC-LEME people studying the geology, geochemistry and landscape evolution of the Puzzler Walls, Charters Towers, in far north Queensland (attempting to use rare-earth-element analyses to see through laterite cover). I have gone on to use the full mix of what I learned at university on a daily basis: geology, chemistry and Indonesian—first, understanding the nature of things, and then applying it on a global scale.

'Of the past 15 years working, only a couple of them have been focused on Australian geology. I have worked mostly on the geochemistry of Ni laterites in many of the tropical areas of Earth: north Queensland, Western Australia, Indonesia, the Philippines, Solomon Islands, Papua New Guinea, New Caledonia, Cuba, Colombia, Turkey and Brazil. In 2008, I worked as the Geology and Mineral Resources Manager for the Pearl Project in eastern Indonesia, on a large Ni-laterite pre-feasibilty study—the pinnacle of a 15-year career. In 2009, I joined the ranks of the unemployed—hopefully just briefly—as our project closed down in Indonesia and Ni laterites became less popular. I plan to change commodities.

'Memorable times at the ANU: bush jumping at Lake Mungo in first year, 1989; memorising all of the rock samples in the second-year lab the night before the practical exam, 1990; late nights spent in the microscope lab polishing thin sections; smashing granites with Professor Bruce Chappell in third year and pondering planetary evolution; glycolating XRD traces with Professor Tony Eggleton and pondering crystal-lattice deformation in honours year, 1992–93; redesigning the ANU shield logo to become a geology society shield volcano [see Plate 8.3].'

Ron Hackney

Ron Hackney graduated with a BSc (Hons) in **1993** and went on to complete an MSc at Victoria University, Wellington (New Zealand), in 1995. The MSc involved geophysical work in Antarctica (a field trip that is yet to be topped) aimed at explaining the origin of the Transantarctic Mountains. After marrying fellow graduate Allison Britt in 1996, Ron followed her to Perth and did a PhD at the University of Western Australia on crustal structure in the Hamersley Province. The PhD, completed in 2001, was followed by a postdoctorate at the Free University of Berlin working on the Andes. At the end of 2004 came baby daughter, Chiara (born on Christmas Day), and a Junior Professorship at the University of Kiel, where, in between teaching, he focused on studies of subduction-zone processes, the application of new-generation satellite gravity data (GRACE, GOCE) and helped initiate high-altitude airborne geophysics on HALO—a new German research jet. Ron returned to Australia in January 2008 to take up a position as a potential-field geophysicist within the Marine and Petroleum Division of Geoscience Australia.

José Ignacio Martínez

'I join from a distance this important celebration. As for my career, this is what I have been doing. After completing my PhD on the palaeoceanography of the Tasman Sea in **1993**, I travelled back to Colombia to take up a position as Exploration Biostratigrapher for the Colombian Petroleum Institute. This lasted for two years; then, late in 1995, I took a position as a Research Fellow back at the ANU, studying the palaeoceanography of the eastern Indian Ocean. Early in 1998, I returned to Colombia—this time to take a position as a Research Lecturer at Universidad EAFIT in Medellin. Since that time, I have investigated the palaeoceanography of the Panama and Colombia basins and the palaeolimnology of Cauca Lake, and directed a number of undergraduate and masters research projects. Some of my students have had the opportunity to take internships in China, Mexico, Chile, Spain and the United States and have participated in French and US cruises. Proudly, three of my former masters students are now doing their PhDs at Bremen, one former undergraduate is doing his PhD at the University of Florida, and others have taken industrial positions. I served as member of the Scientific Steering Committee of Past Climate Changes between 2004 and 2006, and have served on the Editorial Board of Palaeogeography, Palaeoclimatology, Palaeoecology since 1998. Last but not least, gladly we had an ANU student, Max Collett, last semester with us. All these are projections of the Geology Department of the ANU in Colombia. *Hasta la vista—Ignacio.*'

Thierry Corrège

'I completed my PhD with Patrick De Deckker in **1993**, working on the Quaternary palaeoceanography of the Queensland Trough. I then moved to the University of Bordeaux (France), where I had a contract position for two years—

mostly teaching and writing papers from my PhD material. In September 1995, I was hired by ORSTOM, a French Government Research Institute focusing on the tropical zone, with strong interaction with local scientists from tropical countries. ORSTOM changed its name in 2000 and is now called the Institut de Recherche pour le development (IRD). I was first based in Bondy, near Paris, and then moved to New Caledonia in February 1998. I did a lot of fieldwork all over the Pacific Ocean to collect coral samples to study the El Niño–Southern Oscillation through time. Being based in New Caledonia also enabled me to set up new collaborations with colleagues at the ANU (Patrick De Deckker, of course, but also Mike Gagan and Malcolm McCulloch).

'In 2004, I spent a year at the ANU as a Visiting Scientist in RSES. It was really good being back "home", but, after a year, the University of Bordeaux approached me for the position of Professor. I joined Bordeaux in October 2005, first as a Visiting Professor, then as a full-time Professor and Head of the Palaeoclimatology team (about 30 people) since September 2006. My research interest is still on corals and geochemistry to study the El Niño–Southern Oscillation, but I am also starting to work on speleothems from South America to trace the latitudinal movements of the Intertropical Convergence Zone through time.

'Additional details (including many photos) can be found on my research home page: <http://www.epoc.u-bordeaux.fr/indiv/Correge/>'

Roger Skirrow

'Late 1993 to early 1994 was a hectic start to post-PhD life, including submission of the tome, followed by marriage, moving to Adelaide to take up an exploration position with WMC, and the birth of our first child. Fieldwork at Streaky Bay in western South Australia was a pleasant contrast to drilling in the central Gawler Craton in mid-summer. An opportunity to work in Argentina with AGSO took us back to Canberra in 1995, where our second boy was born amid stints of fieldwork on the Palaeozoic metallogeny of the Sierras Pampeanas and struggling with Spanglish. The Argentina project was the last major overseas geoscience mapping project undertaken by AGSO/GA, and I subsequently worked on the Cu-Au metallogeny of the Curnamona Province within the Broken Hill Exploration Initiative.

'From 2000 to 2006, I was Project Leader of the Gawler Craton Project at Geoscience Australia (GA), investigating regional controls on gold and iron-oxide Cu-Au mineral systems. This culminated in the publication of a Special Issue of

Economic Geology; I am also currently on the Editorial Board of this journal. Long-service leave in 2006 took our family to France for three months, where our boys attended a term of school in a village near the Pyrenees. As parents, we had the duty of scouting the local historical sites, cafes and scenic spots to visit later with the boys. En route to Europe, I introduced my boys to the joys and tribulations of mountain climbing when we scaled Mount Kinabalu in Sabah, which, at 4100 m, is a surprising geological phenomenon rising out of the jungles of Borneo. I am currently a Principal Research Scientist at GA, and Leader of the Uranium Systems Project within our new Onshore-Energy Security Program. Basins and regolith are new foci in the program, which is a big shift for most of us Precambrian hard-rockers in the division! We have a great contingent of ex-ANU people here in the division, including Dave Champion, Patrice de Caritat, Anthony Budd and our chief, James Johnson. Sports such as tennis, cross-country and telemark skiing, surfing, hiking and climbing keep me active when they can be squeezed in.'

Greg Miles

'I studied between 1990 and **1994**. My first year was in the "Old" Building, then second year in the "New" Building. I have often recalled how I sort of fell into geology on the advice of an ANU "enrolment councillor" who suggested I take first-year Earth Science as a "filler". At the time, I had morbid plans of studying maths and physics, however, I took only about six weeks to realise my folly, which thankfully coincided with the first Earth Science field trip to the South Coast. It was about this time that I realised that I related better to the natural world and, more importantly, to the people of the department; and that was that! There were still a few hard lessons to learn (most notable was a discussion with old Professor Campbell about why I had scored a miserable 18 per cent on my mid-semester exam; apparently, it had something to do with studying and learning), but, for the most part, my course was set. I have met plenty of other good geos who also fell into the discipline by accident.

'I have plenty of good memories from my geology studies at the ANU. There is no doubt that the various field trips were a highlight, particularly Dave Ellis's trips to Broken Hill [see Plate 4.6], and Camperdown, but even Keith Crook's trips to the South Coast were very good and provided me with a more than fleeting interest in sedimentology (that did not last). Most important were the use of postgraduate students as tutors for labs and field trips such as Leah Moore, Warwick Crowe and Geoff Deacon. These people provided invaluable assistance passing knowledge across what was often a generation gap between lecturer and student.

'I was President of the Student Geological Society in 1992, with the primary function of purchasing beer for the rest of the department's pleasure. The greatest lowlight was being rejected for the Honours Program in 1993 after I did not quite meet the enrolment criteria, and the department would not waive it—bastards! But I got over it and completed a Grad. Dip., which was obviously the right thing to do. I had a job before I even handed my thesis in and promptly moved to Western Australia at the earliest opportunity—in 1994 (before you could realise how poor my thesis was!). I was lucky to be given a gold-exploration job, which really suited me, but I would have taken anything for a first job. I have found the exploration game to be an excellent mix of geological skill, ideas, interpersonal skills and hard work, to mention a few. Deficiencies in one area can be compensated by strengths in others, but a little bit of luck never goes astray either. I am still in Perth (I have spent a bit of time on the east coast, but not much) and these days I am the Exploration Manager for a small junior company called Cazaly Resources dabbling in gold, iron, copper and any other commodity we can make a buck from. It is pretty low profile, but I have done the big-company gig and it does not appeal. I have found far more independence, creativity, autonomy, responsibility and ultimately enjoyment in a small company. And recently I have had a bit to do with John Walshe again in an industry-sponsored research group. Walshey was one of my fourth-year supervisors (along with Dave Ellis) and it has been interesting to listen to his lectures again—it seems like I have come full circle.

Ulrike ('Ulli') Troitzsch

'I came to the department as a visiting overseas student from Germany in **1994** to carry out a field study on the Cooma Granodiorite (honours thesis), working with David Ellis. A highlight of this study was the discovery of hercynite included in sillimanite porphyroblasts—previously unknown for this well-studied locality. Having fallen in love with Canberra (as only visitors from overseas can do) and wanting to do much more mountain biking, I returned to the Geology Department as a PhD student in 1996, to work on 'The crystal structure and thermodynamic properties of titanite solid solution $Ca(Ti,Al)(O,F)OSiO_4$'. This work was of an experimental nature, based on high-PT experiments and the calibration of titanite as a geothermo barometer. This study discovered that pure Al-F end member of titanite $CaAlFSiO_4$—despite its absence in nature—is a stable compound in the laboratory, which initiated a thorough investigation of the crystal structure of this binary titanite solid solution, and study of its thermodynamic properties, in order to explain and reconcile natural occurrences and experimental results (this was awarded a Best

Student Presentation at the 1998 AGU Conference). After a postdoctorate at the University of California, Davis, in 2000, which was very successful in that I met my husband, Patrick, and convinced him that Australia was the place to be, I returned to Canberra and the Geology Department once again—this time to carry out a phase-diagram and crystallographic study of the ZrO_2-TiO_2 system in collaboration with David Ellis and Andrew Christy. This resulted in the discovery of several new phases of special importance for the dielectric ceramic industry, and the publication of several papers and a patent. In 2003, I became a staff member by being appointed Manager of the XRD laboratory and XRF Technician at the department, supporting my research with an FRGS grant and a Knowledge Fund grant at the same time, as well as teaching occasionally. Over the years, I have enjoyed the intimate and collegial atmosphere at the department, and the opportunity to combine my technical skills with research and teaching activities. I am now looking forward to new challenges in the enlarged institution of RSES.'

Tony Meixner

Tony Meixner studied a lot of maths and physics early in his course, so he took several third-year courses as part of his Honours Degree. He was awarded a Hales Scholarship in **1995** to undertake his honours project, supervised by Ted Lilley (RSES) and Peter Gunn (AGSO). He is currently working with the Onshore Energy and Minerals Division at GA, interpreting and modelling potential-field data.

Tony Rathburn

'I was a postgraduate at the ANU working with Patrick De Deckker from 1993 to 1995. After leaving the ANU in 1995, I went to Scripps Institution of Oceanography at La Jolla, California, as a postdoctoral research scientist. From 1999 to 2001, I also became a Lecturer in the Marine and Environmental Studies Program at the University of San Diego. In 2001, I joined the Geology Program at Indiana State University, where I am currently an Associate Professor, and a Research Associate Scientist at Scripps Institution of Oceanography.

'Although based on another continent, I have continued collaborative work with Patrick De Deckker and others at the ANU, including participation on the initial AUSCAN cruise with Patrick. Throughout my career, my research has focused on the ecology and bio-geochemistry of benthic foraminifera. My work includes the use of submersibles to study sea-floor methane-seep environments and submarine canyons off California and Alaska. I am also working on foraminiferal responses to pollutants in the Venice Lagoon in Italy, and recently have returned to the Antarctic to examine sea-floor ecosystem responses to temporal changes in ocean productivity.'

Tim Barrows

Tim Barrows graduated with First-Class Honours from the Geology Department in **1995**. 'My honours mapping indicated the important possibility of glacial permafrost deposits on Black Mountain in Canberra. I won a few awards in my undergraduate degree [see Prize Lists]. I then received a John Conrad Jaeger Scholarship and completed a PhD at the Research School of Earth Sciences in 2000. I also received the Robert Hill Memorial Prize in that year. My first Postdoctoral Fellowship was with the Geology Department, jointly with the University of Colorado. My second postdoctorate was for two years in the Nuclear Physics Department at the ANU, where I was promoted to Research Fellow. My third postdoctorate was an ARC APD—also conducted at the Nuclear Physics Department at the ANU. Most of my research has centered on climate change during the Quaternary. I have specialised in marine micropalaeontology and cosmogenic nuclides for dating.'

David Tilley

In 1991, David Tilley received a BSc (Hons) with a major in Geology from Flinders University of South Australia and, in **1996**, a PhD from the ANU in Regolith Mineralogy. His research thesis was on the evolution of bauxitic pisoliths at Weipa in north Queensland, which resulted in the discovery of the natural occurrence of eta-alumina in bauxite. After his PhD, David continued to work at the ANU's Geology Department, as an Associate Lecturer in Geology and Mineralogy, while conducting research into poorly diffracting materials within the Australian regolith. This included studies into the alteration of titanite, the weathering of iron–nickel meteorites and the characterisation of hisingerite using advanced X-ray diffraction techniques and electron microscopy. After seven years with the ANU, David worked in the Australian Government in resources and energy policy for a period of five years. Here he assisted in drafting legislation, ministerial correspondence and briefing for the Minister for Industry, Tourism and Resources. David returned to geology in 2006 as an Exploration Geologist with Oceana Gold Limited in New Zealand, where he supervised diamond drilling and geochemical sampling operations and conducted a study on the use of the portable infrared mineral analyser in gold exploration. At the end of 2007, David joined Amdel Limited in Adelaide as a

Senior Mineralogist and was the manager of Amdel's new QEMSCAN laboratory. In July 2008, David began working with Archer Exploration Limited, where he is currently an Exploration Geologist.

Leanne Dancie/Armand

Leanne Dancie/Armand (**1991–97**), honours student, Research Assistant and PhD graduate, joined the Geology Department from Flinders University, Adelaide, via a stint at the Alice Springs Museum, to work as a part-time Research Assistant in vertebrate palaeontology for Professor David Ride. At the same time, she undertook a part-time Honours Degree (perhaps the first in the Geology Department) studying the geology and palaeontology of Teapot Creek in the Monaro region of New South Wales, under the supervision of Professor David Ride and Professor Ken Campbell (1991–92). She achieved a First-Class Honours and applied for several PhD scholarships, settling on a new focus in marine micropalaeontology with Patrick De Deckker in the Geology Department, and Dr Jean-Jacques Pichon (deceased) at the University of Bordeaux I. For this thesis (1993–97), she developed the first palaeo-sea ice reconstruction of the Southern Ocean based on the remains of fossilised diatoms (micro-algae) from deep-sea cores recovered south of Tasmania. During the final year of her thesis, Leanne migrated to France in the hope of attracting a postdoctorate in the northern hemisphere.

Several months later, in mid-1998, she found herself a postdoctorate in Tasmania, at the Institute of Antarctic and Southern Ocean Studies (IASOS), and returned to Australia with her family (husband and three-month-old son). Here her research remained focused on the Antarctic-diatom records covering the past 120 000 years. A subsequent fellowship in the Antarctic CRC followed and allowed her to branch into the living world of Southern Ocean phytoplankton. At the demise of the Antarctic CRC in 2003, she was employed for a year at CSIRO Marine Laboratories as a Research Assistant while awaiting an eventually successful outcome in the prestigious European Union's Marie Curie Postdoctoral Fellow Awards in 2005. She thus returned to France to take up the Fellowship at the Laboratory of Oceanography and Biogeochemistry at the Centre of Oceanography of Marseille (CNRS/Universite de la Mediterranee) focusing on issues of phytoplankton distributions in relation to the uptake of silica and the role of iron availability in the Southern Ocean. In 2007, Leanne was awarded the Australian Academy of Science's Dorothy Hill Award for outstanding women scientists in the fields of palaeontology, taxonomy, geology and oceanography and the Rose Provasoli Scholarship from Bigelow Laboratories, USA.

Leanne is now back in Tasmania in her reintegration phase of the Marie Curie Fellowship co-funded by CSIRO Marine and Atmospheric Research and the Antarctic Climate and Ecosystem Research CRC at the University of Tasmania, where she continues her diatom work crosscutting the fields of modern and palaeoceanographic research under the classical domains of biology, geology, oceanography and sea-ice cover. She has had a second son and is now fluent in French.

Leah Moore

After leaving the ANU in **1997**, Leah Moore undertook a variety of positions with the University of Canberra—first, with the School of Resource, Environment and Heritage Science as Lecturer in Regolith Geology and Volcanic Sedimentology, and then as Convener of second-year Earth Sciences. She was promoted to Senior Lecturer in 2001 and became Director of the Dryland Salinity Hazard Mitigation Program, and Course Manager for the Graduate Certificate in Physics Teaching in the School of Information Science and Engineering. She then progressed in administrative roles, becoming Dean of Students in 2004, Deputy Head of Division (2006–08) and Acting Pro-Vice Chancellor, Communication and Education (2007). In 2005, she took study leave in curriculum studies with the Faculty of Education, University of British Columbia, Vancouver. Leah is married, with two young girls, and is currently working in Perth and Canberra.

Kriton Glenn

Kriton Glenn graduated with honours in **1997** and started contract work for Environment Australia to assist in the management plan for development and enforcement of the *Environment Protection and Biodiversity Act* (1999) onsite at Ashmore Reef. He provides both linguistic and scientific support with advice on the environmental challenges in managing the offshore marine reserves on the North-West Shelf. This involves working closely with the Australian Customs National Marine Unit, the Royal Australian Navy and a range of scientific institutions. He is currently completing a part-time PhD at Adelaide University.

Shawn Stanley

'After completing my Honours Degree in **1997**, I worked as a Research Assistant to examine desert-quartz grains in the alpine lakes of the Snowy Mountains. I then joined GA, where I am currently involved in Australia's submission to the UN Commission on the Limits of the Continental Shelf in support of our claim for extended continental shelf under the UN Convention on the Law of the Sea. I am currently working in the field of Geographic Information Systems and I am creating a computer interface to help display Australia's law of the sea claim to the United Nations, and have provided GIS support in the finalisation of our claim areas. I have also continued sedimentation studies off eastern Antarctica. The well-rounded Earth Science Degree I gained from the ANU has proved invaluable in furthering my career.'

Dave McPherson

'After graduating in **1998** and failing an interview with Shell, I was offered a position in the three-year Graduate Program at Woodside Energy in Perth as an exploration geologist. This seemed a better proposition than "fly-in–fly-out" mining work, although I had little background in petroleum. The second half of 1998 proved to be life changing for me: in July, my son was born; in September, I was offered the job with Woodside; in November, I completed my honours thesis; in December, I was married and moved to Perth to start our new life there.

'During that first year, I was introduced to basin modelling, stratigraphic interpretation, seismic interpretation, and drilling operations (including visiting an offshore drilling rig for a few weeks to learn about the "pointy end" of the oil business). The next few years flew by in the Geologic Services team, where I quickly became involved in some of Woodside's high-profile projects, including exploring around the Laminaria oil field in the Timor Sea, developing a sequence-stratigraphic model south of the Goodwyn and Rankin gas fields, and performing pore-pressure predictions in a number of sedimentary basins, including the Gulf of Mexico and offshore Mauritania, in west Africa. For three years, I was involved in Woodside's exploration program in the Gulf of Mexico, based in Perth. This team had everything I wanted at the time: exciting geology, an active exploration program, a challenging business environment and the opportunity to travel to Houston. I became closely involved in the drilling of wells in the deepwater Gulf of Mexico, with water depths ranging from 1000 to more than 3000 m. This required more and more frequent travel to the United States, and eventually led to the decision to relocate to Houston with my family. In mid-2003, we packed up the contents of our newly built house in Perth and boarded the plane. For my wife and children, this was to be their first overseas trip, so they were very excited but also nervous about what lay ahead.

'In 2006, I decided to leave Woodside to take up a position with Shell in Houston, where I was involved in a very large project to acquire new acreage in the deepwater Gulf of Mexico beneath large salt bodies that made seeing the structure beneath using seismic data very difficult. I worked in a multidisciplinary team evaluating this acreage, determining its economic value and risk, deciding how best to extract the oil and gas, and then, finally, how much to pay for the leases. The culmination of this was the federal lease sale held in October 2007 in New Orleans, where we (Shell) secured some of the most prospective leases in the sale.

'With the end of such a large and exciting project came a sense of anticlimax—and in me a desire to find the next challenge. I did not have to look far to find it: Shell was busy setting up an exploration team in Nigeria to target deep, high-pressure oil and gas pools that lie below the existing prolific fields. This was the challenge I was after—interesting geology and a very different way of living—so, in July 2008, we packed up again and began our African adventure.

'When I look back on my journey so far, I can honestly say that the oil industry has gone beyond my expectations—formed back when I first applied to Shell. I am doing very interesting and challenging work, using my geologic skills daily in an industry that touches the lives of everyone on the planet. Although many of the skills I use today were taught on the job, the broad background I gained at the ANU has been invaluable in helping me achieve my goals.'

Sophie O'Dwyer

Sophie O'Dwyer graduated from the ANU in **1999** with a BA and BSc, and then took a Graduate Certificate and MA in Environmental Management and Development (2000–01). After graduation, she worked in Chiang Mai, Thailand, on environmental education, and at the World Resources Institute in Washington, DC, on landfill-gas projects. After returning to Australia, in 2003, she worked at the Brisbane City Council on landfill-gas projects and water and energy-efficient community projects. She transferred to the Queensland Environmental Protection Agency in 2004 to manage the Renewable Energy Diesel Replacement Scheme, which oversaw the installation of solar power in off-grid areas of Queensland. She worked closely with the local renewable energy industry to promote their sustainable growth and quality assurance. Currently, she is Managing Director of Commerce Carbon Group, an energy solutions consultancy and a world leader in carbon management services, where she specialises in greenhouse-gas management plans, emissions inventories, carbon offset and brokerage services.

Heather Catchpole

'I graduated with a BSc majoring in Geology at the ANU in **2000**. I then moved to Tasmania briefly with my partner, Cody Horgan—also a Geology graduate. We met studying geology and shared some wonderful field trips; we have been together for 11 years and have been married for four years. We have two kids, Saskia (five) and Beren (10 months). While Cody studied a boat-building degree in Tasmania, I returned to Canberra and worked as a Research Assistant at the Australian Defence Force Academy and did a Masters in Science Communication at the ANU. I worked for a while at *The Canberra Times*, and then at the CSIRO, where for three years I was Editor of the children's science magazine *Scientriffic*. Cody moved back to Canberra, bringing his first 12-ft huon-pine dinghy, and we had our daughter, Saskia. We then moved to Sydney, where I worked as a science journalist for the ABC Science News site and Cody worked as a shipwright. I worked at the ABC for the next three years as a journalist and science-web producer, writing articles, creating web-based science games and working on projects such as the plastic bag famine for National Science Week. In 2007, I enrolled in a Certificate of Fine Arts and became pregnant with our second child, Beren. I now work from home as a freelance science writer and production editor of the Geological Society of Australia's magazine, *The Australian Geologist*.'

Lynda Radke

'I completed a PhD in **2000** in the Geology Department at the ANU, under the direction of Dr Patrick De Deckker. The work, which examined the relationship between the geochemical evolution of salt lakes and ostracod-species composition, was an extension of earlier research carried out in Canadian groundwater, spring and stream systems. I started to work at GA in 1998, researching estuary bio-geochemistry in the Urban Coastal Impacts Group and subsequently the Coastal CRC. My current role at GA is to manage the content of the Oz Coasts web site (formerly, Oz Estuaries), and to investigate the potential for geochemical surrogates of marine biodiversity.'

Cameron O'Neill

Cameron O'Neill graduated with a First-Class Honous Degree in Geology and a BEc in **2000**. He is currently working as an analyst with the Australian consulting firm ACIL Tasman, providing economic and strategic analysis to underpin policy decisions by government and providing advice to industry leaders and financial institutions to help them anticipate these policy outcomes. Current areas of research include water pricing and allocation, energy-market structure, and prospective greenhouse policy and its impact on business. 'I find this type of work diverse and challenging and it enables me to utilise the skills I learnt in both disciplines at the ANU.'

Michelle Spooner

Michelle Spooner graduated with BSc (Hons) in **2001** and a PhD in **2006**. 'Since leaving the ANU, I have worked in the Marine and Coastal Environment Group at GA, where I focused on the biodiversity of rhodolith deposits in the Recherche Archipelago. I am currently working in the domestic carbon capture and storage group.'

Kurt Worden

Kurt Worden graduated with BSc (Hons) in Geology and LLB degrees in **2001**. He joined GA and has been trained as a geochronologist, specialising in U-Pb SHRIMP dating, which is carried out at RSES. He works with the NT Geological Survey participating in the planning and fieldwork for their geochronology program.

Samantha Williams

Samantha Williams gained an Honours Degree in **2002** for work on an area of the Tianshan Mountains in north-western China. She is now working as a geologist at the Cowal Gold Project in New South Wales for Barrick Gold of Australia. Her work involves statistically evaluating the accuracy and precision of the gold-assay results to determine the economic viability of the project. She also does core logging, interpretation, report writing and database validation. In the future, she plans to do a PhD in petrology and tectonics.

Davide Murgese

Davide Murgese came from Turin, Italy, to gain his PhD in **2003**. He studied the evolution of the eastern Indian Ocean over the past 60 000 years by analysing the carbon and oxygen stable isotopes in calcareous micro-organisms on the sea floor. One of the major goals of this research was to gain information on the changes in carbon dioxide levels for the Quaternary period to understand, and possibly predict, the greenhouse effect and global warming. He is now back in Turin, working for the Environmental Division of CSI Piemonte as a geologist assessing environmental impacts.

Peter Collett

'I graduated in **2006** and have been working as an exploration geologist for the past two years—initially, with Rio Tinto Exploration and, recently, with Tri Origin Minerals based at the Woodlawn mine site. I have largely kept abreast of changes in the department since I left, and continue to bear an interest in the state of geoscience education at the ANU and more broadly within Australia. It would be a fantastic opportunity to be involved in celebrations of the ANU Geology fiftieth anniversary both to have the opportunity to meet alumni and to gauge the impact that graduates have had on Australian mining, exploration and geoscience.'

Thomas Abraham-James

'Upon successful completion of my BSc (Hons) in **2006**, I began full-time employment with Rio Tinto at their Argyle diamond mine. I relocated to live in Kununurra, Western Australia, and was given the role of Mine Geologist and member of Rio Tinto's graduate program. My primary role was to assist with Argyle's underground development, being actively involved in all geological and geotechnical matters.

'After a year at Argyle, I made the decision to follow my heart and deviate to a career as an exploration geologist. I gained employment as Project Geologist with Platina Resources Limited, a platinum-group-element (PGE) explorer based in Queensland. This achievement was assisted by the PGE experience that I gained during my time at the ANU, having won the AusIMM Southern Africa–Australia scholarship. This enabled me to travel to South Africa for my honours project, conducting research at Placer Dome Incorporated's Sedibelo PGE project. Since joining Platina, I have operated at all of the company's projects, located in Western Australia, Namibia and Greenland. I have now been with Platina for two years and, as of December 2008, was promoted to the position of Exploration Manager.'

Appendix 1

Visiting Fellows

1962
Prof. P. Ramdohr	Heidelburg Univ.
Prof. D. Williams	Royal School Mines UK

1963
Dr Maxwell Gage	Canterbury NZ
Prof. B.H. Mason	Columbia New York
Dr M. Schwarzback	Univ. Cologne
Prof. W Berry	Univ. California Berkeley
Dr P. Harris	Univ Leeds UK
Prof. S.K. Runcorn	Univ Newcastle UK
Dr J.B. Waterhouse	NZ Geol. Survey

1964
Prof. F.H.T. Rhodes	Univ. College Swansea
Prof. H.K. Charlesworth	Univ. Alberta
Prof. N. Bogdanov	Univ. Moscow

1965
Dr J.T. Kingma	NZ Geol. Survey
Prof. R.L. Folk	Univ. Texas
Dr Shohei Banno	Univ. Tokyo
Prof. Tuzo Wilson	Univ. Toronto
Prof. F.J. Turner	Univ. California Berkeley
Prof. A. Volborth	Nevada School Mines
Prof. T. Barth	Univ. Oslo
Prof. A. Boucot	California. Inst. Techn.
George Grindley	NZ Geol. Survey

1967
V.B. Olenin	Moscow State Univ.
Dr H. Jaeger	Humbolt Univ. E. Berlin
Dr D. Skevington	Univ. College London
Dr K.S. Thomson	Yale University
Prof. J.B. Waterhouse	Univ. Toronto
Prof. R. Weiner	Colorado School Mines

1967 - continued
Prof. B.H. Mason*	US Nat. Museum Washington
Prof. W.B.N. Berry	Univ. California Berkeley
Dr I. Rolfe	Hunterian Museum, Glasgow
Dr H. Brunton	BM (Nat. Hist.) London
Dr G.A. Chinner	Cambridge Univ
Prof. W.W. Moorhouse	Univ Toronto
Prof. F.A. Campbell	Univ. Calgary Alberta
Prof. J.B. Lerbekno	Univ. Alberta Edmonton

1969
Dr M.S. Krishnan	Geol. Survey India
Dr R.O. Brunnschweiler	UN. Development
Dr A.P. Subramaniam	New Delhi
Dr R.W. Willett	DSIR, NZ

1970
Prof. W. Burnham	Pennsylvania State Univ.
Prof. H.B. Whittington	Cambridge Univ.
Prof. V.T. Frolov	Moscow State Univ.
Dr D.R. Bowes	Univ. Glasgow
Prof. A. J. Naldrett	Univ. Toronto
Dr R.P. Suggate	NZ Geol. Survey
Prof. F.B. Conselman	Texas Tech Univ.
Dr J.H. Guillon	Union Minieres, Noumea

1971
Dr A. Sugimura	Univ. Tokyo
Prof. A.S. Romer	Univ. Harvard
Dr G.R. Stevens	NZ Geol. Survey
Dr. R.A. Cooper	NZ Geol. Survey
Dr H. Okada	Kagoshima Univ. Japan

Geology at ANU (1959-2009)

1972		**1981 - continued**	
Dr J. Ferguson	Univ. Witwatersrand, South Africa	Dr P.J. Stephenson*	JCUNQ
Dr N. Sobolev	Academgorodok. USSR	Dr C. Pollard	JCUNQ
Prof C.J. Hughes	Memorial Univ. Canada	Dr C. Johnston.	JCUNQ
Prof. V.V. Drushchits	Dpt Palaeontology Moscow State Univ.	**1982**	
		Dr D. Haynes	Western Mining
Prof. P. Sonnenfeld	Univ. Windsor, Ontario Canada	Dr R. Hine*	BHP
		1983	
1973		Dr M. Smith*	Dental Anatomy, Guys Hospital, Lond.
Prof. F.F. Langford	Univ. Sascatchewan, Canada		
		Dr E. Stevens	St Andrews Univ. Scotland
Dr Yu. S. Borodaev	Moscow State Univ.		
Prof. L.A. Frakes	Univ. Florida	Prof. S. Guggenheim*	Univ. Illinois
Prof. J. Rodgers	Yale Univ.	R. Patterson	Ardlethan
Dr J.F. Truswell	Univ. Witwatersrand.	M. Jones & J. Angus	Goldfields
1975-77	*None listed*	**1984**	
1978		Dr P.J. Hancock*	Univ. Bristol UK
Prof. C.Y. Craig	Univ. Edinburgh	Dr T. Laska	Univ. Calgary
Prof. L.T. Silver*	California Institute Technology	Prof. W. Manser	Univ. PNG
		Prof. D. Raup	Univ. Chicago
Prof. F. A. Frey	*Institution not recorded*	Dr K. Smith	NSW Instit. Technology
1980		Prof. R. Stanton*	Univ. New England
Prof. D. Wones	Virginia State Polytechnic	**1985**	
		Prof. N. James	Memorial Univ. Newfoundland
Dr I.A. Smith*	Univ. Auckland NZ		
Dr N. Arndt	Saskatoon	Prof. B. Oaks	Utah State Univ.
K. Norrish* & J. Hutton	CSIRO Soils, Adelaide	Prof. C. Barker	Univ. Tulsa USA
W.A. Watters	NZ Geol. Survey	Sir Charles & Lady Fleming	Wellington NZ
1981			
Delegation	Academia Sinica	Prof. Ohmoto	Pennsylvania State Univ.
Prof. E.H.T. Whitten	Univ. Chicago	Dr C. Heinrich	AGSO
Prof. W.S. Pitcher	Univ. Liverpool UK	Group of 6	Univ. Mining & Metallurgy Cracow.
Dr Y. Zheng	Beijing Univ. PR China		
Prof. R. Walker	McMaster Univ. Canada	**1986**	
Dr C. Kendall	Gulf Research & Development USA	Prof. R. Fox*	Univ. Alberta
		Prof. M. Stauffer	Univ. Saskatchewan
Prof. G. Lister	Univ. Utrecht Holland	Prof. H.P. Schultz	Univ. Kansas
Prof. D. Merriam	Syracuse Univ. USA	Dr P.J. Coleman	Univ. WA
Prof. P. Cloud	Univ. Santa Barbara	Prof. A. Niem & Dr Niem	*Institution not recorded*
Dr K. H. Wolf	Consultant Sydney	Prof. P. Roehl	Trinity Univ. Texas
Prof. A. Salvador	Univ. Texas		
Prof. K. Hsu	ETH Zurich		

Appendix 1

1987
R. Ruddock	Univ. Auckland NZ
Prof. M. Obata	Kumamoto Univ. Japan
Prof A. Knoll	Harvard Univ.
Dr T. Barber	Royal Holloway London
Dr D. Lewis	Waite Agricultural Instit.
Prof. E. Nisbet	Univ. Saskatchewan
P. Wells*	Victoria Univ. Wellington
Prof. A.J.R. White*	Latrobe Univ. (retired)
Prof. W.D.L. Ride*	Univ. Canbera (retired)

1988
Dr P.J. Coleman	Univ. WA
Dr W.M. Last	Univ. Manitoba Canada
Dr. S. McKerrow	Oxford Univ.
Dr M. Obata	Univ. Kumamato Japan
Dr Z Zhou	Xi-an Institute China
Dr P. Vail	Esso Lecturer
M. Rubenach	JCUNQ
Dr A. Ewart	Univ. Queensland
A. Urbaneek	Polish Academy Science
P. Rodda	Geol. Survey Fiji
P. Warren	Royal Soc. London

1989
Dr M. Andrews	Royal Scottish Museum
Dr S. Ehara	Kyushu Univ. Japan
Prof. I. Lerche	Univ. South Carolina
Dr J. Lowell	Esso Lecturer
Prof. D. Nahon	Univ. Aix-Marseille
Prof. J. Parrish	Univ. Arizona
Dr W. Shitao	Inst. Geol. Sci. Beijing
Group 6	Energy Develop Org. Tokyo Japan
Prof. L.P. Zonenshain	Instit. Oceanography, Moscow
Wu Yangqiang	Shanxi Mining Coll.

1990
Prof. P. Bankwitz	Potsdam GDR
Prof. J. Berg	N. Illinois Univ USA
Prof. B.M. Funnell	Univ. E. Anglia, UK
Prof. Y. Hiroi	Chiba Univ Japan
Prof. Y. Ming-zhen	Computer Sci. PRC
Dr E. Purdy	Esso Lecturer

1990 - continued
Prof. J. Temple	Birkbeck Coll. London
Dr F. VanderHor	Manus Basin project
Prof. J. Wang	Xian Petrol. Inst. PRC
Dr W. Yim	Honk Kong Univ.
Prof. G. Zuffa	Univ. Bologna Italy
Dr. G. M. Taylor*	Univ Canberra
Dr M. Ayress	ANU/Univ. Aberystwith
Dr N. Williams*	BMR
Dr N. F. Exon*	BMR
Prof. P. Brown*	St. Andrews Univ Scotland
Prof. H. Okada	Yamagata Univ. Japan
Prof. E-an Zen	US Geol. Survey
Dr W. D. Turnbull	Field Museum Chicago
Prof. W. Pinxian	Tongi Univ. Shanghai
Prof. R. C. Whatley	UCW, Aberystwyth

1992
Dr R.E. Barwick*	ANU Zoology (retired)
Dr K.A.W. Crook*	Univ. Hawaii
Dr Q. Bone	Editor Phil. Transac. Royal Soc. Lond.
Dr H. Bjerring	Mus. Nat. Hist. Sweden

1993
Prof. P. Candela	Univ. Maryland
Prof K.S.W. Campbell*	ANU (retired)
Prof. Y. Hiroi*	Chiba Univ. Japan
Prof. J-J Pichon	CNRS. Univ. Bordeau
C. Hiramatsu	JAPEX Japan
Prof. R.W.R. Rutland*	AGSO
Dr K. Shiraishi	Polar Res. Japan
Dr P. Ahlberg	Zoology Mus. Oxford.
Dr W. Hamilton	US Geol. Survey
H. Neil	Waikato Univ. NZ

1994
Dr J.A. Clack	Cambridge University
Prof. R.C. Fox	Univ. Alberta
Dr S. McMillan	Crown Res. Inst. NZ
Dr I. Metcalfe*	Univ. New England
Prof. P. Wang	Tongji Univ. Shanghai
Dr S. Umino	Shizuoka Univ. Japan
Prof. P.W. Candela*	Univ. Maryland
G. Morison	Klondyke Exploration

Geology at ANU (1959-2009)

1995		1998 - continued	
Dr R. Ahmed*	ANU	Dr. G.H. Cameron	ANU
Dr. C. Ballhaus	Max Planck Instit.	Dr P Carr*	Univ. Wollongong
Dr R.W. Johnson	AGSO	Dr C. Carson	Univ. Sydney
Dr S.G. McMillan	Crown Res. Dunedin	Dr J. Casey*	AGSO (retired)
Dr K. Shiraishi	Nat. Inst. Polar Res. Japan	Dr K.K .Singhal	India
		Dr E. Truswell*	AGSO (retired)
Dr S. Umino	Shizuokai Univ	T. Wagner	Univ. PNG
Dr J.L. Walshe	CSIRO	Dr S. A. Welch	Univ. Wisconsin USA
Dr G. Warren	AGSO (retired)	Prof. C. Shaoping	Seismological Bur. Beijing
Dr P. Ahlberg	Brit. Mus. Nat. History		
Dr M. Ayress	AGSO/ANU	J. Gamble	Victoria Univ. NZ
N. Krupina	Inst. Vert. Palaeo Moscow	Dr P .Hancock	CRES
		P. Jones *	AGSO (retired)
Dr O. Lebedev	Inst. Vert. Palaeo Moscow	Dr I. Lindley	Yass NSW
		Dr W. Mayer*	Univ. Canberra (retired)
Dr E.Mark-Kurik	Eastonia Acad. Sci.	Dr. C.L. Moore	ANU
Dr S. Nees	Geomer. Kiel, Germany	H. Patia	Rabul Volcano.Observ.
1996		Prof. W. Ranson	Furman Univ. South Carolina USA
Dr H.J. Harrington*	AGSO (retired)		
Dr M. Kawano	Kagoshima Univ. Japan	1999	
Dr R.W. Johnson*	AGSO	Dr G. Chaproniere*	AGSO (retired)
Dr S. Okamura	Hokkaido Education Univ.	Prof. S. Erickson	Virginia Poly. Inst. USA
Dr P.J. Potts	Open Univ. UK	Prof. B.R. Frost	Univ. Wyoming
Dr J.L. Walshe	CSIRO	Dr J. Gillespie	
Dr G. Warren	AGSO (retired)	Prof. S. Eriksson	Virginia Poly. Institute
Dr Y. Martynov	?	Dr J. Knutson	AGSO (retired)
Dr D. Wyborn*	GA	Dr M. Le Gleuher*	ANU
Dr J. Knutson	GA	Dr D.E. Mackenzie*	AGSO (retired)
Dr J. Sheraton	GA	L. Maeda	Tohoku Univ. Japan
Dr I. Parkinson	ANU	Dr W. Mayer*	Univ. Canberra (retired)
Dr T. Esat	RSES	Dr R.S. Nicoll*	AGSO (retired)
Dr C. Allen	ANU	J.A.B. Palmer	Divs Computing ACT
Dr S.E. Eggins	RSES	H. Patia	Rabul Volc. Obs.
1997		Dr Shafik	AGSO (reired)
Dr J.M. Caton *	Archaeology & Anthropology ANU	Dr J.W. Sheraton	AGSO (retired)
		Dr S-S Sun*	AGSO
Prof. M.A. Velbel	Michigan Sate Univ.		
Dr G. Young*	AGSO (retired)		
Dr D. Strusz*	AGSO (retired)		
1998			
Prof. J. Banfield*	Univ. Wisconsin		
Dr R. Burne*	AGSO (retired)		

Appendix 1

2000

Prof. R. Frost	Univ. Wyoming
Dr S. Ericksson	Virginia Poly. Institute
Prof. L. Zhang	Beijing Univ. PRC
Prof. Jouzel	French Polar Institute
Dr C. Klootwijk*	AGSO (retired)
Prof. S.R. Taylor*	RSES (retired)
Dr P. Blevin*	AGSO
R.J. Bultitude	Dept. Mines & Energy Qsld
Dr P.F. Carr	Univ. Wollongong
Prof. T.H. Green	Macquarie Univ.
Dr J. Knutson	AGSO (retired)
L .Maeda	Tokoku Univ.
Dr M.T.G. de Oliveira	Brazil
Prof. C .Ollier	CRES
H. Patia	Rabaul Volc. Observ. PNG
Dr S. Shafik	?
Dr C. Tarlowski	AGSO

2001

T. Barly	
Prof. X.Y. Chen	
Prof. R.A.Eggleton*	ANU (retired)
Dr C.L. Moore	Univ. Canberra.
Dr J. Palmer	
C. Parvey	AGSO (retired)
J.K. Payne	
F. Reith	Germany

2002

Dr J. Knutson	AGSO (retired)
Dr P. Lennox	Univ. NSW
Dr T. Ulrich	Univ. Qsld
Dr M. Baltuck	NASA
Dr R. Binns*	CSIRO (retired)
Dr J. Clarke	ANU
Dr J. Giddings	AGSO (retired)
J. Payne	AUSLIG (retired)
Dr M. Revel	Univ. Grenoble
Dr L. Shaffi	Univ. Parma/Univ Cambridge
H. Stagg*	GA
J. Wilcox	GA

2003

Dr A-M. Boullier	CNRS Grenoble France
Dr J. Knutson	GA (retired)
Dr M. Revel	Univ. Grenoble France
Dr T. Sen	
Dr M.J. Rickard*	ANU (retired)

2004

Dr F. Gingele	Baltic Sea Res Instit.
Dr T. Kawakami	Okayama Univ. Japan
Dr. E. Perkins	Alberta Res. Council
K. Scott	CSIRO
Prof. S. Yang	Univ. Geosci. Wuhan PRC

2005

Y. Chen	China Univ. Beijing PRC
Dr E. A. Felton*	Univ. Hawaii (retired)
Dr A. Glikson	RSES
Dr J. Long*	Sci. Mus. Victoria
Prof. H. J. Marchant*	Antarctic Divs /CSIRO
Dr T. Haymet	CSIRO
C.H. Parvey	(GIS) ANU
K. Scott	CSIRO

2006

Prof. K. Kyser	Queens Univ Canada
Dr A. Nutman	RSES

2007

Dr B. Fordham	CSIRO
Prof. C. Von der Borch	Formerly Marine Geol. Flinders Univ
P. Hill	GA (retired)
Dr E. Perkins	Alberta Res. Council

2008 (current Visiting Fellows at EMS & #working in RSES)

T.T. Barrows#	ANU
R. Burne	AGSO (retired)
K.S.W. Campbell	ANU
J. Chappell#	ANU
W. Compston#	ANU
K.A.W. Crook	ANU
R. Eggleton	ANU
N. Exon	AGSO (retired)
E. A. Felton	ANU
B. Fordham #	

2008 (current Visiting Fellows at EMS & #working in RSES) - continued	
A. Glikson	AGSO (retired)
D. Green#	ANU
C. Klootwijk	AGSO (retired)
K. Lambeck#	ANU
T. Lilley#	ANU
J. Magee	ANU
H. Marchant,	CSIRO
I. McDougall#	ANU
M. Paterson,#	ANU
M. Rickard	ANU
R. Rutland,#	AGSO
K. Scott	CSIRO
D. Strusz	AGSO (retired)
R.Taylor	ANU
P. Treble	
E. Truswell,	AGSO/ANU
S. Turner#	ANU
S. Welch#	ANU

* long-term Fellows or those who have made multiple visits
\# working in RSES

Appendix 2

List of BSc Honours Theses

Year	Student	Supervisor	Title
1963	Lambert, B.	AJRW	The geology of the Berridale district, NSW.
1965	Black, L. P.	AJRW	The geology of the Eucumbene area.
1967	Carr, M. J.	KAWC	The Skillion: a study of cyclic landscape evolution.
	Coventry, R. J.	KAWC	The geology of Shingle House Creek valley, NSW, with particular reference to its Cainozoic history.
	Raine, J. I.	KAWC	Geology of the Nerriga area with special reference to Tertiary stratigraphy and palynology.
	Richards, D. N. G.	MJR	Geology of the Jerangle district.
	Walraven, F.	RAE	The geology of Milton-Conjola district.
1968	Bein, J.	KAWC/MJR	Geology of the Tantangara area, NSW.
	Collings, P. S.	AJRW/KLW	The geology of the Tarago area, NSW.
	Legg, D. P.	MJR	The geology of the Cooleman Plains and surrounding areas, NSW.
	Mackenzie, D. E.	AJRW	Basalts and metamorphic rocks of the Kiandra area, NSW.
	Patterson, K. R.	AJRW	Geology of the southern end of the Berridale Batholith.
	Roberts, D. E.	AJRW	The geology of Gunning-Cullerin area: a petrological and chemical study of the Gunning granite and its contact aureole.
	Torr, G. L.	AJRW/AEG	Geology of the Wheeo-Biala area, NSW.
1969	Arndt, N. T.	RAE	The igneous and metamorphic rocks of the Walwa area, north-eastern Victoria.
	Christie, D. M.	JAMcD	Geology and geochemisty of Crown Reef, Norseman, Western Australia.
	Curtis, L. W.	BWC	Geochemistry of high-potassium rocks, southeastern NSW.
	Davoren, J.	KSWC	The geology of the Taralga-Bannaby district, NSW.
	Felton, E. A.	KAWC	Geology of the Mount Fairy area, NSW.
	Huleatt, M. B.		The geology of Paleozoic sediments south-east of Taralga, NSW.
	Labutis, V.	KAWC	The geology of the Yarrangobilly area, NSW.

Geology at ANU (1959–2009)

Year	Student	Supervisor	Title
	Mckay, W. J.	AJRW	Metamorphic and igneous rocks in the Tallangatta district, north-east Victoria.
	Powell, I. L. L.	KAWC	The geology of the Eurobodalla area, NSW.
	Price, R. C.	RAE	Granites of the north-east Victorian metamorphic complex.
	Williams, N.	AJRW	Paleozoic geology of Bethana goldfield area, NE Victoria.
	Young, C.	KSWC	The geology of the Burrinjuck–Wee Jasper area, NSW.
1970	Baczynski, N. R. P.	MJR	Geology east of Bredbo, NSW.
	Bucknell, W. R.	MJR/KAWC	Environment of deposition and deformation of the Merrimbula group, south of Eden, NSW.
	Kelly, G. R.	KAWC	Geology and soils of the Towrang district, NSW.
	Madsen, P. N.	AJRW	Palaeozoic rocks of the Cowra district, NSW.
	Moignard, P. S.	KSWC	Geology of the Boambolo district, NSW.
	Nisbet, B. W.	MJR	Structural geology of the Captains Flat area, NSW.
	Veijayaratnam, M.	AJRW/BWC	Petrology of the Myalla Road syenite complex and its environment.
1971	Gibson, D. L.	KAWC	The solid and surficial geology of the catchment of the Upper Genoa River, NSW and Victoria.
	Gorter, J. D.	DAB	The geology of a portion of Cooper and Nubrigyn parishes, near Stuart Town, NSW.
	Griffin, T. J.	BWC	The Moruya Complex, NSW.
	Hine, R.	BWC/AJRW	Granite studies of the Kosciusko Batholith, Jindabyne, NSW.
	Hughes, R. J.	KSWC	The geology of the Jerrawa–Coolalie area, NSW.
	Wain, P. J.	MJR	The deformational history of the Mallacoota area, Victoria.
1972	Bell, M. W.	KSWC	The geology of the Jemalong Range area in central NSW.
	Duff, B. A.	MJR	The structural geology of the Arltunga region, central Australia.
	Gardner, C. M.	RAE	The geology of syenites at Mittagong, NSW.
	Holloway, D. J.	KSWC	Part I. The geology of the Mundoonen Range, NSW. Part II. The fauna of the Riverside Formations (upper Silurian) of Canberra.
	Jeffrey, D. G.	MJR	Geology of the Kiandra area, NSW.
	Nazer, R.	KSWC	The geology of the Trundle–Bogan Gate district, NSW.
	Palmer, K.	MJR	The geochemistry and tectonic evolution of the Cooleman volcanic centre, NSW.
	Sadler, D.	JAMcD	The geology of Cadia, NSW.
	Scott, P. A.	KSWC	Devonian stratigraphy and sedimentology of the Araganui area, NSW.

Appendix 2

Year	Student	Supervisor	Title
	Tassell, C. B.	KSWC	The geology of Quidong, NSW.
	Topp, G. C.	JAMcD	The geology of the Rye Park area.
1973	Cameron, J.	BWC	Granites of the Bombala area, NSW.
	Creaser, P. H.	CEBC	The geology of the Goulburn-Brayton-Bungonia area, NSW.
	Gibson, B. G.	JAMcD/MJR	The geology of the Bigga-Tuena area, NSW.
	O'Brien, K J.	JAMcD	The depositional environment of Captains Flat-type mineralisation.
	Vincent, C. S.	BWC	The geology of the Nimmitabel-Brown Mountain area, NSW.
	Williams, C. S.	RAE/IAS	Geology and geochemistry of the Happy Jacks-Jagungal area.
	Williams, I. S.	BWC	Granites north of Jindabyne, NSW.
1974	Atkins, B. N.	KAWC	The geology of the Minjary district.
	Clark, J. M.	RAE/IES	Geology and geochemistry north east of Coolac.
	Colwell, J. B.	CEBC	Geology of the Warrumba Range area, NSW.
	Drake, G. W.	JAMcD	Geology of an area east of Purnamooda, Broken Hill, NSW.
	Gibbings, S. A.	RAE	The geology of the Coolac-Gobarralong district.
	Kennard, J. M.	KAWC	The geology of the Brungle district.
	Pillans, B. J.	KAWC	Surficial geology of the Murrumbidgee-Bredbo Interfluve.
1975	Beams, S. D.	RAE	The geology and geochemistry of the Wyndham-Whipstick area.
	Coleman, A. R.	KSWC	Geology of the Murga area, NSW.
	Cooke, J. A.	JAMcD	The geology of an area north of Boorowa, NSW.
	Hill, R. I.	BWC	The geology of an area east of Michelago, NSW.
	Lesh, R. H.	RAE	Geology and geochemistry of the Candelo-Bega region, NSW.
	Levy, I. W.	JAMcD	Geology of the Bluebush area, Western Australia.
	Stevens, M. K.	KAWC	The geology of the south Gundagai district, NSW.
	Williams, A.	KSWC	Geology of the Parkes-Bumberry region, NSW.
1976	Bomford, R.	KAWC	Yarrangobilly stratigraphy.
	Creaser, H. J.	DAB	Part I. Geology north of Windellama.
		KSWC	Part II. A study of the cleithrolepis fauna of the Sydney Basin.
	Kennedy, A. B.	DAB/KSWC	Geology of the Capertee Valley.
	Lowde, B. J.	RAE/JAMcD	Part I. Processing of mineral assemblage images.
			Part II. A study of slope stability in a close-jointed rock mass.

Geology at ANU (1959–2009)

Year	Student	Supervisor	Title
	Rigden, S.	RAE	Lockhart: variations on a mineralogical theme.
	Webber, R. B.	BWC	A geological study of the Bethungra region, NSW.
1977	Berents, H. W.	JAMcD	Geology of the Neville–Barry region, NSW.
	Collins, W. J.	BWC	Geology and geochemistry of the Gabo Island granite suite and associated rocks.
	Goleby, B. W.	MJR	The geology of the Majors Creek district, NSW.
1978	Rixon, L. K.	MJR	Structural features of the upper Devonian of the South Coast of NSW.
1979	Johnston, R.	BWC	Geology of Yass-Binalong area, NSW.
	Mangold, G. R.	KAWC	The geology of the Darbalara area.
	McDonald, K. A.	BWC	Geology of the Burrinjuck area, NSW.
	Slater, R.	MJR	Structural analysis and exhalite studies in the Peelwood area, NSW.
	Ward, P. G.	KAWC	Some geological insights into the Noojee-Tooronga area, west Gippsland.
1980	Barrett, A. G.	RAE	The geology of an area southwest of Wombat, NSW.
	Habelko, P. F.	KAWC	The geology of the Califat area near Tumut, NSW, with a two-dimensional gravity profile across the Tumut Trough, NSW.
	Morrison, F.	BWC	The geology of syenites at Benambra, northeast Victoria.
1981	Banfield, J. F.	RAE	The geology of the Murrabaine Mountain.
	Compston, D.	MJR	Aspects of the geology of the Tumblong area, NSW. Research report included: report of a gravity traverse across the Gilmore fault zone near Tumblong, NSW (1980).
	Davidson, J.	WEC	A contribution to the geology of the Mt Wright area, NSW.
	Ferguson, I.	T. Lilley/ KSWC	Telluric current investigations near Tumut, NSW.
	Gray, M.	WEC	The geology of an area south of Rockley.
	Platts, W. D.	RAE	The geology and geochemistry of a granite-gabbro association south of Cowra, NSW. Bound with: Geology of Kuroko deposits: a review.
	Ralser, S.	MJR	Palaeozoic geology of the Benanderah area, northeast of Batemans Bay, NSW.
	Tulip, J.	RAE	The mystery of Mount Dromedary, with special reference to Little Dromedary.
	Wheller, G. E.	WEC	The geology of the Aranbanga region, Queensland.
1982	Killick, C.	KAWC	Sedimentology and structural geology of Bumbolee Creek Formation, Tumut Trough.
	Taylor, G. R.	KAWC/GMT	Landscape evolution of Lake Bunyan and environs, Cooma district.

Appendix 2

Year	Student	Supervisor	Title
1983	Boston, A.	JLW	Alteration and gold mineralisation with the Boyd volcanic complex, South Coast, NSW.
	Konecny, S.	BWC/WEC	Geology and geochemistry of the Gunning granites and geology of the Lionel ultramafic complex with emphasis on the possibility of finding a diamond-source region.
	Robertson, N.	KAWC	Sedimentary aspects of the Worange Formation, South Coast, NSW.
	Rogers, J.	JLW	Sundown tin prospect, New England.
	Turner, B.	BWC	Geology and geochemistry of the Eugowra granitic complex.
1984	Alexander, G.	KAWC	Upper Molonglo regional surficial geology.
	Bird, M.	JLW	Geology and geochemistry of Briars mine and environs.
	Creaser, R.	JLW	Some geological aspects of the Anduramba region, southeast Queensland.
	Jones, M. J.	KSWC	Aspects of the geology from the Mount Yambira area.
	Olley, J.	MJR	Geology of Primrose Valley area, NSW.
1985	Arthur, J. F.	JCT	Geology of an area near Beaudesert in the Clarence-Moreton Basin, Qld.
	Flindell, P. A.	JLW	Geology of the Kookynie area, WA, with emphasis on gold mineralisation.
	Graham, B. J.	JLW	Geology, geochemistry and mineralisation of an area near Majors' Creek, NSW.
	McCormack, C.	BWC	Oberon granites.
	Palmer, S.	JCT	Geology of an area near the Tamrookun, Clarence-Moreton Basin, Qld.
	Raymond, O. L.	JLW	Geology and mineralisation of Mt Wright area, N. Qld.
	Smithies, R. H.	JLW	The geology and mineralisation of the Burrandana area.
1986	Voorhoeve, H.	KAWC	Problems in fold belt geology of the Wattle Flat–Limekilns district.
	Wilkinson, C.	KAWC	Seismic investigations in the eastern Manus Basin, Papua New Guinea.
1987	Conan-Davies, M.	RAE/MJR	Part I. A sheet silicate and fluid inclusion study of the mine area D. N. W., Olympic Dam.
			Part II. Geology Tathra volcanics.
	Sheppard, S.	JLW	Geology and geochemistry of a gold-bismuth-tellurium prospect.
	Urbaniak, S.	DJE	Geology of the Heathcote–Pyalong area, central Victoria.
	Yacopetti, C. M.	BWC	Geology of the area south of Wyangala Dam, NSW.
1988	English, P.	DJE	Geology of Mt Bogong and Mt Wills, Victoria.

Geology at ANU (1959–2009)

Year	Student	Supervisor	Title
	Grant, M.	JLW	Geology of Junction Reefs, and Sheahan-Grants, gold deposit, Mandurama, NSW.
	Jagodzinski, K.	DJE	Geology of the Tatong area, Victoria.
	Newman, C.	JLW	The geology and mineralisation of the south west zone, Rosehall Caldera, south east Queensland.
1989	Bygrave, S.	JLW	Geology of Burnt Yards area and mineralisation of the Glendale deposit, Mandurama, NSW.
	Coad, D.	DL	Lacustrine depositional environments of the northern Eyre Peninsula, South Australia.
	Rumble, C.	RAE	Geology of Lake Tyrrell region, north west Victoria, with emphasis on heavy minerals.
	Vicary, M.	BWC	Geology of the Killiecrantie area.
1990	Deacon, G. L.	PDD	Biostratigraphy of the Clifton Formation, Otway Basin, Victoria.
	Gordon, M.	JLW	Ordovician volcanics of the Wellington area, NSW.
	Crowe, W.	MJR	A study of a Mylonite zone along the eastern margin of the Murrumbidgee Batholith: the structure, microfabric development and evolution.
	Stanner, J.	RAE/B. Lees	Late Quaternary geomorphology and development of laterite/bauxite at Gove, NT. [Geography thesis]
	Mills, M.	JLW	The geology of the Navilawa area, SPL 1218, Viti Levu, Fiji.
	Hope, M.	JLW	Part A: The geology of the Errowanbang area.
			Part B: The mineralogical zonation across the Junction Reefs gold deposits, Mandurama, NSW.
	Woodhouse, W.	DJE	Tourmalinites from Yanco Glen: a low-grade metamorphic region of the Broken Hill Block.
	Schubert, C.	JCT	The stratigraphic signature of sea-level changes in the Shoalhaven Group of the Sydney Basin.
1991	Boda, S.	JLW	The geology, structural setting and genesis of the Chester Mine, northwest Tasmania.
	Findlay, C.	KSWC	Vertebrate fossils, facies and palaeoenvironment of the Upper Taemas Subgroup, Yass, NSW.
	Shaw, J.	DJE	The geology and mineralisation of an area east of Cooma and the Peak View Outlier, NSW.
	Franklin, C.	KSWC/DJE	Part I. The upper Devonian Hervey group near Grenfell, NSW: palaeoenvironment and structural history.
			Part II. The marine environment of Prydz Bay, Antarctica: microbiota and facies distribution.
	Jones, B.	JLW	Geological setting and genesis of the Endeavour 44 Au-Pb-Zn skarn, Parkes, NSW.
	Goody, A.	B. Kennet	Broad-band studies of the upper mantle beneath northern Australia.

Appendix 2

Year	Student	Supervisor	Title
	Barry, R.	PDD	The neodymium, strontium and rare earth element analyses of conodonts from the early Ordovician Horn Valley siltstone, Amadeus Basin, Australia.
	Fellows, M.	JCT	A stratigraphic and palaeo-environmental study of the Bungendore Formation, Lake George, NSW.
1992	Corkery, R.	RAE/TL	Thin-sheet modelling of the Australian continental crust.
	Shepherd, A.	MJR	The structure and stratigraphy of the Tooram Hills, near Cobar, NSW.
	Edgecombe, S.	DJE	The geochemistry and geological setting of a suite of spinel-bearing xenoliths from the Monaro Volcanic Province, southeastern NSW.
	Dimmer, L.	PDD	Geology and evolution of lacustrine sediments within the newer volcanics, Beeac, Western Plains District, Victoria.
	Coyle, K.	RAE	The formation of kaolinite: alkaline Earth and transition-metal halide intercalation complexes and their derivatives.
	Love, S.	MJR	Part I. Possible ages and origins for some rocks of the mid-Proterozoic Willyama Supergroup, NSW.
			Part II. Timing of kink folding in the Wapengo–Bunga area, South Coast, NSW.
	King, P.	BWC	A-type granites from the Lachlan Fold Belt: a case study and reassessment.
	Williams, S.	KAWC	Part I. Origin, formation and subsidence history of seamounts in the Christmas Island offshore region, Indian Ocean.
		N. Exon	Part II. A processing sequence for high resolution seismic data collected over deep water, by R/V Rig seismic, using high-speed towing with a short cable.
	Girvan, S.	JLW	Geology and mineralisation of the Copper Hill porphyry copper-gold deposit, near Molong, NSW.
	Budd, A.	JLW	Internal differentiation and mineralisation of the Banshea Pluton, NSW.
	Dansie, L.	D. Ride	Stratigraphy and palaeontology of Teapot Creek, Sherwood and Boco stations, near Nimmitabel, NSW.
1993	Britt, A.	RAE	Regolith, bedrock and geochemistry at Johnson Creek, northwest QLD.
	Hackney, R.	MJR/R. Griffiths (*RSES*)	The behaviour of a migrating viscous sheet descending to a density or viscosity interface and some comparisons with the Earth's mantle.
	Rivers, C.	RAE	Regolith development, geochemistry and landscape evolution of the Puzzler Walls, Charters Towers, north Queensland.

Geology at ANU (1959–2009)

Year	Student	Supervisor	Title
	Witham, B.	JLW	Structure and vein formation associated with the Inglewood Fault, west of Scotland mine, Gympie, QLD.
1994	Barr, K.	JLW/P. Blevin	The geology and mineralisation around Dairy Hill, NSW.
	Carey, M.	RSES	Petrography and geochemistry of selected sills from the Kambalda–Kalgoorlie region, WA.
	Harvey, N.	JLW	The relative timing of mineralisation and alteration, with respect to the structure of the Glendale Deposit, NSW.
	Patrick, K.	RSES	Part I. The partitioning behaviour of phosphorus in the upper mantle under subsolidus and hypersolidus conditions.
			Part II. High pressure melting experiments on altered oceanic basalt—implications for the petrogenesis of modern adakites.
	Woodbury, M.	JLW	Red Dome & Mungana porphyry Cu-Au and base metal skarns of NE QLD.
1995	Holgate, F.	DJE	The geology of the Great Serpentinite Belt Upper Bingara, NSW.
	Hudson, T.	BNO	Recharge to the Coonamble embayment in the Great Artesian Basin, NSW.
	Kovacs, G.		Alteration and mineralisation in relation to structure at the Triangle Park Prospect.
	Barrows, T.	PDD	Three aspects of late Quaternary palaeo-climatic reconstruction in eastern Australia.
	Meixner, T.	T. Lilley/MJR	Aeromagnetic modelling of the Bonaparte Basin.
	Fein, M.	RAE	Basalt weathering formation near Taralga, NSW.
1996	Devlin, C.	DJE	Evolution of the Broken Hill albitites: evidence from the Stirling Vale Syncline.
	Adamson, S.	RAE	Exploration geochemistry and regolith stratigraphy of the Beechmore Block, Parkes, NSW.
	Burkle, P.	PDD	Aspects of the dissolution and ecology of planktonic foraminifera in the eastern Indian Ocean, radiolarians and benthic foraminifera.
	Foster, L.	RAE	Sedimentary reworking across the Weipa bauxite deposit.
	Wilson, M.	BNO	Ngilgi Cave—the history of cave development, Yallingup, WA.
	Smith, B.	RAE	Calcretes; their nature and association with gold and other metals. Evidence from Broken Hill and Parkes, NSW.
1997	Read, C.	RJA	The Goalen Head Gabbro: a mafic pluton in the Lachlan Fold Belt, Australia.
	Spry, M.	RAE	Tertiary processes on the coastal lowlands, Brooman and Tuross, south east NSW.

Appendix 2

Year	Student	Supervisor	Title
	Cairns, B.	PLB	Geology of the Adelong goldfield.
	Chorley, M.	BWC	Geochemistry and geochronology of the Tenterfield granites.
	Crawford, M.	RAE	Geology, mineralogy, geochemistry and landscape evolution of the Burdekin Downs/Fletcherview area, north Queensland.
	Darby, L.	BNO	Geological evolution of the Scott Plateau since the late Cretaceous, with discussion on palaeoceanography and sea level.
	Glenn, K.	BNO	Aspects of the geological development of Ashmore Reef, North West Shelf, Australia.
	Hack, A.	JM	Structural development of a turbidite-hosted gold (-bismuth) quartz vein deposit: Union Hill mine, Maldon, central Victoria.
	James, M.	BWC	The geology of the Gullengambel area, central New South Wales.
	McIntosh, I.	DJE	Experimental studies in the system PbS-FeS-ZnS-(Ag2S) at 1 atmosphere to 27 kilobars: implications for the Broken Hill orebody.
	Pottenger, D.	RAE	The nature and distribution of the regolith in the Bulong area, Western Australia, and its implications for gold exploration.
	Pulford, A.	Vander Beek/ RSES	Cenozoic landscape evolution of the east Australian highlands: constraints from Miocene basalts, Blue Mountains, NSW.
	Spandler, C.	RJA	The petrology and mineralisation of the Greenhills Ultramafic Complex, Southland, New Zealand.
	Stanley, S.	PDD	Aeolian input to the sedimentation of Blue Lake, New South Wales.
	Warren, K.	PDD	Late Quaternary history of Lakes Eyre and Clayton: a micropalaeontological study.
1998	Holzapfel, M.	RAE	Regolith–landscape evolution and silcrete investigation: Redon East, Broken Hill Block, NSW.
	McPherson, D.	PDD	Late Pleistocene and modern periglacial landforms of Mt Ginini, ACT/NSW border.
	Ryan, D.	PDD	Holocene production and accumulation of sediments on Wistari Reef, southern Great Barrier Reef Province: the significance of coral reefs as global CO_2 sources.
	Shirtliff, G.	RAE	Massive gypsum, ferricretes and regolith landform mapping of Western Balaclava, Broken Hill, NSW.
	Spandler, M.	PLB	The geology of the Mineral Hill Field, central NSW: igneous evolution and Cu/Au mineralisation.
1999	Alexander, J.	SFC	The structure and geochemistry of the Inrepide and Santa Ana gold mines, Kambalda, Western Australia.

Geology at ANU (1959-2009)

Year	Student	Supervisor	Title
	Belfield, S.	RAE	The 1.8ka Taupo eruption: nature and origin of geochemical variations in pyroclastic products.
	Burch, G.	PLB/JM	The geology and mineralization of the Mount Mackenzie region, central southeastern Queensland.
	Buttefield, G.	RAE	The regolith, soil and sediment geochemistry of the Mount Dromedary Igneous Complex.
	Dempsey, D.	PLB	Cargo porphyry copper-gold deposit, NSW, petrography, alteration and geochemistry.
	Douglas, N.	DJE	An experimental study of bismuth (and gold) in the hydrothermal environment.
	Gunton, C.	JM	A study of molybdenum at the Ernest Henry Cu-Au deposit, northwest Queensland.
	Hackney, D.	JM/H. O'Neil RSES	The effects of Cr on partial melting in the CMAS system.
	Ison, M.	SFC	Structural setting and evolution of the Schmitz and Skinner nickel sulphide orebodies, Tramways Belt, Kambalda, WA.
	Jain, P.	*RSES*	The origin of gahnite at Broken Hill, NSW.
	Kositcin, N.	PLB	The petrology and geochemistry of the Avenall Basic Intrusive Complex, southeastern New South Wales.
	Lacey, R.	PNC	Integration of the geology and geophysics of the Yeoval Complex, NSW.
	Lee, S. Y.	RAE	The geochemistry and mineralogy of the Marlborough nickel laterite.
	Leslie, C.	RAE	High-resolution seismic imaging of paleo-channels near West Wyalong, NSW.
	Mitchell, C.	JM	Controls on development of gold mineralization at Challenger gold deposit, Adelong, NSW.
	O'Neill, C.	RJA	Cumulate rocks of the Luxmore Complex, Fiordland, New Zealand.
	Richardson, M.	W. Taylor	Upper mantle processes, diamond-bearing potential and stratigraphy of the mantle in the Kimberley region, Western Australia: evidence from chromite, chromian diopside and garnet xenocryst chemistry.
	Ross, R.	RAE	The sources of suspended sediment and associated phosphorus for Adelong Creek catchment.
2000	Farmer, C.	PNC	Studies of seismic source characterisation with the neighbourhood algorithm.
	Fitt, M.	RAE	Regolith evolution, Gidginbung, NSW.
	Fokker, M.	RJA	The petrology of the Jurassic dolerite sill, Mt Geryon, Tasmania.
	Howe, B.	JM	3-D modelling of the Sedibelo Exploration Tenement, Bushveld Igneous Complex, South Africa.

Appendix 2

Year	Student	Supervisor	Title
	Isaacs, D.	PLB	Evolution of the Nymagee region: utilising geochemistry, geochronology and geophysics.
	Johnson, A.	JM/PLB	A study of the trace-element and included-mineral composition of molybdenite.
	Kilby, M.	RAE	Sources and sinks of sulfur at Ranger Uranium Mine, NT.
	Nicholson, C.	SFC	Fluid wallrock interaction and structural processes in the North Orchin gold lode, St Ives Gold Operations, Kambalda, WA.
	Sloan, M.	JM	Relationships between the ore and enclosing rocks at Broken Hill, NSW.
2001	Anderson, G.	RAE	The influence of macro- and meso-biota on regolith development and evolution in a dry sclerophyll forest.
	Dunn, A.	JM	Chalcophile element partitioning during granite intrusion.
	Earl, K.	RJA	Geochemical variation in the late Quaternary eruptive products of the Taupo volcano fractionation.
	O'Leary, R.	SFC	Dolerite-hosted gold mineralisation at the Argo mine, St Ives Gold Operations, WA.
	Page, T.	PNC	An integrated study of the geophysics, structure and geochemistry of the Tarana granite, Bathurst, NSW.
	Sinclair, N.	AGSO	The sedimentology, palynology and environment of the deposition of the mid to late Triassic (Ladinian to Norian) of the Challis Oil Field, Northwest Shelf, Australia.
	Spooner, M.	PDD	The late Quaternary palaeoceanography of the Banda Sea east of Timor with implications for past monsoonal climates.
	Worden, K.	RJA	Petrology of the Takitimu Mountains, Southland, New Zealand.
	Anderson, I.	J. Clark	Stratigraphy, regolith and landforms in the Balladonia region, Western Australia.
2002	Craven, C.	RJA	Trace-element geochemistry of charnockites.
	Crowther, B.	PNC	A remote sensing approach to regolith mapping: an example from the Cootamundra area, NSW.
	Dalby, K.	RJA	Trace-element behaviour in peralkaline rhyolites: a study of melt inclusions from Mayor Island (Tuhua), New Zealand.
	Day, R.	JM	Nature of hydrothermal alteration at Goonumbla, NSW.
	Dowell, K.	JM	Origin of black opal nobbies in Lightning Ridge, NSW, Australia.
	Gretton, E.	PDD	Palaeo-oceanographic changes offshore New Caledonia for the past 140,000 years.

Geology at ANU (1959–2009)

Year	Student	Supervisor	Title
	Kalinowski, A.	JM	An experimental investigation into the causes and effects of sulfide partial melting at Broken Hill, NSW, Australia.
	Mapham, B.	RJA	The petrology and geochemistry of the West Bismarck Arc, Papua New Guinea.
	Milczyarak, T.	*RSES*	Receiver functions and the structure of the mantle below Australia.
	O'Callaghan, S.	DJE/JM	The origins of albitisation and pyrite mineralisation, Broken Hill, Australia.
	Proko, K.	RSES	Investigation of transfer functions between seismograms.
	Williams, S.	DJE	High pressure low temperature metamorphism and geotectonic evolution of the South Tianshan (Xinjiang, NW China).
2003	Bewert, K.	BMP/KMcQ	A regolith landform approach to environmental management in the Cadia mining area, central NSW.
	Francis, J.	KSWC	Depositional environment, palaeontology and taphonomy of the Hatchery Creek Formation, NSW.
	Glanville, H.	KMcQ	Regolith-landform mapping, leucitite basalt and the landscape evolution of the Byrock region, northwestern NSW.
	Lilly, K.	*RSES*	Extreme El Nino events recorded in Great Barrier Reef corals.
	Tynan, S.	BNO	Dissolution of shallow water carbonate sediments as a potential sink of atmospheric CO_2.
	Webb, J.		The role of shrink-swell soils: an investigation of the Fowlers Gap patterned ground.
	Worthy, M.	J. Magee	*No copy*
	Wykes, J.	JM	Just add H_2O—hydrous sulfide melting in the FeS-PbS-ZnS-H_2O system at 1.5 GPa.
	Zasiadczyk, S.	SFC	Structural controls and gold mineralization in the North Revenge area, Kambalda–St Ives, WA.
2004	Bamford, P.	KMcQ	Geochemical dispersion and under-cover expression of gold mineralisation at the Wyoming gold deposit, Tomingley, NSW.
	Bishop, K.	Lister *RSES*	The FΔSZ conundrum: microstructural analysis of the Faros Tectonic Slice, the Island of Sifnos, Aegean Sea, Greece.
	Clulow, J.	Lister *RSES*	Structural geology of the South Vanoise Massif, French Western Alps.
	Gregory, C.	Lister *RSES*	Microstructural analysis of the Main Central Thrust, NW Himalaya (India).
	Hamilton, T.		Sources and sinks of iron and manganese in the Cotter River catchment, ACT.

Appendix 2

Year	Student	Supervisor	Title
	Kehoe, M.	S. Welch	Investigating the role of biotic verses abiotic processes in acid sulfate soils in coastal NSW.
	Kuhnen, M.	PDD	Constraining the source areas and nutrient transport of sediments entering the Fitzroy Estuary since European arrival.
	Matic, V.	PNC	Integrating geophysics for regolith mapping in central Queensland.
	Munro, D.	KMcQ	Regolith controls on geochemical dispersion in the CSA area, Cobar, New South Wales.
	Reed, A.	G. Lister *RSES*	Structural analysis of the Faros Tectonic Slice, the Island of Sifnos, Aegean Sea, Greece.
	Richardson, L.	BNO	An interpretation of a Sr/Ca record in a coral growing at 20m water depth.
	Riesz, A.	D. Kriste	The origin and mobilisation of salts in the Hovells Creek catchment, NSW.
	Rogers, J.	PDD	Radiolarian assemblages in surface sediments of the eastern Indian Ocean adjacent to Australia.
	Southby, C.	M.Gagan *RSES*/BNO	Late Holocene El Nino variations recorded in fossil corals from Papua New Guinea.
	Stanton, L.	BNO	Structure and evolution of the margin of Eastern Wilkes Land and Terre Adelie, Antarctica.
	Stevens, M.	JM	Constraining platreef petrogenesis, northern limb of the Bushveld Complex, South Africa.
	Summerhayes, E.	BMP	Zinc mobility in the regolith: hemimorphite solubility.
	Tailby, N.	JM	Evolution of the Spitskop carbonatite.
	Woolrych, T.	KMcQ	Regolith–landforms of the Cobar gold field and geochemical dispersion at the Illewong Prospect.
2005	Abraham, J. T.	RJA	Evolution and mineralisation of the Bushveld Complex: evidence from ultramafic pegmatites.
	Ayling, M.	S. Eggins *RSES*	Taphonomy and geochronology of the Holocene Sub-Fossil Estuarine Assemblage, Robe, South Australia.
	Beavis, F.	S. Welch	The mobilisation of anthropogenic lead in estuarine environments.
	De Livera, J.	JM	Copper mobility in iron-rich regolith: reactive transport experiments.
	Edwards, J.	BNO	Palaeo-oceanography of the Timor Sea during Marine Isotopic Stage 3.
	Fry, N.	JM	The solubility of Ag in supercritical hydrothermal brines.
	Hunt, J.	G. Young	An examination of stratigraphy and vertebrate fish fauna of the middle Devonian age from the Hatchery Creek Formation, Wee Jasper, New South Wales.
	Mikkeslon, N.	RJA	The major element geochemistry and oxidation states of back-arc basins in the South West Pacific.

Geology at ANU (1959–2009)

Year	Student	Supervisor	Title
2006	Briggs, E.	BNO	Shoreface sand supply onto beaches, Moruya Beach, NSW.
	Bambic, L.		*No copy*
	Bonato, P.	GA	Fuel-moisture content mapping in the ACT region prior to the 2003 Canberra bushfires.
	Byrne, H.	BNO	Assessing recent decadal climate change through geochemical analysis of coral cores from the Timor Sea.
	Collett, P.	JM	The Rossing uranium deposit, Namibia—geochemistry and geochronology of a granite-hosted uranium deposit.
	Hickey, A.	SFC	The petrological zonation, fractionation and alteration of the condenser dolerite, Argo gold mine, WA.
	Higgins, A.	S. Welch	Physicochemical properties of sediments, the Loveday Disposal Basin, South Australia.
	Hui, S.	BMcP	Sources and sinks of uranium in the Albury–Wodonga region.
	Isaacson, L.	S. Welch	The distribution of acidity, salts and metals in a highly modified acid sulfate soil backswamp, Mays Swamp, Kempsey, NSW.
	Kinnison, K.	S. Welch	Chemistry of flood waters in a waterway affected by acid sulfate soils, Kempsey, NSW.
2007	Bermingham, K.	RJA	Eucrites: an investigation into early planetary differentiation.
	Hughes, J.	B. Pillans *RSES*/ I. Roach	Regolith geochronology and landscape evolution of the Wombat area, NSW.
	Kelly, T.	R. Grun *RSES*	Strontium-isotope tracing in animal teeth at the Neanderthal site of Les Pradelles, Charante, France.
	Owens, R.	BNO	History of sea-surface temperature and carbonate chemistry from the Timor Sea during the last interglacial.
	Roberts, J.		*No copy*
	Williams, S.		*No copy*
	Yates, G.	S. Welch	Salinity dynamics in a sand bed stream; the Wollombi Brook, Upper Hunter catchment, NSW.
	Cockburn, A.	S. Beavis/S. Welch	Exploring relationships between the spatial variability of aqueous geochemistry and the distribution of frogs in an acid-sulfate soil landscape on the mid-north coast of NSW.
	Newton-Walters, D.	G. Young	An examination of vertebrate fish remains (Arthrodire placoderms, Dipnoan osteichthyes, and acanthodians) from the early Devonian of Burrinjuck, NSW, Australia.

Appendix 3

List of Graduate Diploma Theses

Year	Student	Supervisor	Title
1973	Anderson, G. R.	MJR	The geology of the Frampton area, NSW.
	Brush, J.	CEBC	The geology of the Ornmier area, NSW.
	Clarke, R. A.	?	Geology of the Nungar Range area, NSW.
	Goldsmith, R. C. M.	MJR	The geology of the Lacmalac area, NSW.
1974	Fortowski, D.	MJR	The geology of an area north-west of Cootamundra, NSW.
	Hansen, T.	JMcD	Geology of an area west of Forbes, NSW.
	Roberts, P.	MJR	Geology of an area west of Cootamundra, NSW.
1975	Knight, G. F.	MJR	The geology of an area east of Bega, NSW.
	Nicholson, P. O.	MJR	The geomagnetics and dykes of the Central Tilba area, NSW.
	Percival, P. J.	RAE	Geology and geochemistry of the Cobargo area, NSW.
1976	Moorhouse, P. R.	KSWC	The structural geology of the Mandagery Syncline and its regional significance.
1977	Gooday, P.	MJR	Geology and geochemistry of the Kybeyan area, NSW.
1982	Schmidt-Mumm, A.	WEC	The geology of an area northwest of Rockley.
1983	Wu, Q.	KSWC	Facies analysis of Cavan limestone and the associated rocks.
1985	Gordon, I.	MJR	The Palaeozoic structural geology of the Nelligen area near Batemans Bay, NSW.
	Hakim, S. M.	MJR	The geology of the Fairlight area, NSW.
	Sharman, P.	KSWC	Palynology of Goat Paddock Crater, WA.
1986	Szabo, L.	MJR	Lachlan Fold Belt fault patterns.
	Veitch, S.	KAWC/GMT	Landscape evolution of Cathcart area, NSW.
1987	Ringwood, K.	DJE	The crystallisation of iron-rich olivines in natural rocks and synthetic analogue systems.
1988	de Vries, R.	JLW	Geology of EL2642, the Barry, NSW & Geology of the Browns Creek gold mine, Blayney, NSW.
	Jankowski, P. E.	JCT	The geology of the Caroda-Rocky Creek area, NSW.
	Langford, R. P.	KAWC/KSWC	Cainozoic palaeogeographic evolution of the Australian continental margin. [2 vols]

Geology at ANU (1959–2009)

Year	Student	Supervisor	Title
	Maconachie, L.	RAE	The mineralogy and evolution of a basalt weathering profile, Berridale, NSW.
	Wilford, J.	KAWC/B. Lees	Use of remote sensing data in detecting possible surface alteration associated with leaking hydrocarbons over the Palm Valley gas field, central Australia.
1989	Hawkes, G. E.	RAE	The Cainozoic history and hydrology of the Lake Bathurst catchment, NSW.
	Prihardjo, S.	MJR	Palaeography of Kirawin Formation, Taemas, NSW.
1990	Miller, C. R.	RAE	Geology, geochemistry and mineralization of an area surrounding the Michelago iron mine, NSW.
1991	Clausen, L.	KAWC	Fluvial architecture in the upper Devonian Worange Point Formation, Merimbula Headland, north of Eden, NSW. Presentation of data, discussion of methods and preliminary interpretations.
	Mendosa, A.	JCT	Geophysical investigation of the Menindee and Blantyre troughs in the Murray-Darling Basin, western NSW.
1994	Miles, G.	DJE	The geology and mineralization of the Lewis Ponds-Icely districts, NSW.
	Kendon, B.	RAE	The effects of European settlement on the catchment and upper basin of Burrill Lake, NSW; a comparative study.
1995	Troitzsch, U.	DJE	A petrological study of the high-grade metamorphic rocks of the Cooma Complex, NSW.
2002	Bean, L.	KSWC	*Cavenderichthys talbragarensis*; a transitional fish of the Kimmeridigan from Talbragar, NSW.
2004	Jones, R.	D. Kirste	Aspects of regolith and salinity in the Cowra area: a comparision between weathering regimes.
	Rawsthorn, K.	PNC	The effect of fracture geometry and fluid flow on the thermal performance of the hot dry rocks system in the Cooper Basin, South Australia: 3D numerical modelling.
	Gerner, E.	PNC	A geostatistical approach to mapping the temperature in the Australian crust.
2006	Alorbi, A.		The geology and geochemistry of the Minadetta area near Nyngan, NSW.

Appendix 4

List of MSc and MPhil Theses

Year	Student	Supervisor	Title
1964	Jensen, A. R.	KAWC	The geology of the Blenheim area, Bowen Basin, Qld.
	Johnson, N. E. A.	DAB/KAWC	The geology of the Krawaree area, NSW.
	Wilson, E. G.	KAWC/MJR	The geology of the Farrar–Hoskinstown area, NSW.
1965	Mollan, R. G.	AJRW/BWC	Tertiary volcanics in the Peak Range, central Queensland.
	Smith, J. W.	MJR	The geology of the Gundaroo-Nanima area, NSW.
1966	Cook, P. J.	KAWC	The Stairway sandstone—a sedimentological study.
	Jones, P. J.	DAB	Upper Devonian ostracoda from the Bonaparte Gulf Basin.
	Rhodes, J. M.	AJRW	The structure and chemistry of feldspars in selected Australian granites.
1970	Halford, G. E.	AJRW	Dykes and their inclusions from Kelly's Point, NSW.
1972	England, R. N.	AJRW	Progressive metamorphism of amphibolites from the Cloncurry and Petermann ranges.
	Koluzs, P. J.	DAB	Sedimentology and environmental reconstruction of the Cavan limestone, Taemas district, NSW.
1975	Kurylowicz, L. E.	CEBC	Basin analysis and petroleum prospects of the Larapinta group, Amadeus Basin, with particular reference to porosity evaluation.
1976	Davoren, P. J.	KSWC	Distribution of Permian brachiopods.
	Shields, L.	KSWC	The ostracode taxonomy and palaeoecology of the Receptaculites limestone, Taemas near Yass, NSW.
1977	Lightner, J. D.	KAWC	The stratigraphy, structure and depositional history of the Tumut region, NSW.
	Creaser, P. H.	KAWC	Lithogenesis and diagenetic features of recent and ancient terrigenous limestones.
	Ramli, N.	JCT	Permian sedimentology of the Jervis Bay district, ACT & NSW.
	Watchman, A. L.	RAE	Contact metamorphism of the Heathcote greenstone at Mount William and Mount Carmel, central Victoria.
1978	Henry, R. L.	KAWC	Stratigraphy and structure of Tarago district, NSW.
	Close, R. J.	JLW	The geology of the Wyangala–Lyndhurst region, central NSW.

Geology at ANU (1959–2009)

Year	Student	Supervisor	Title
1979	Browne, P.	KSWC	Graptolite palaeontology and biostratigraphy, Snowy River, NSW.
	McConnell, A.	KAWC/GT	A landscape history of the Wantabadgery area, NSW.
1980	Hardjoprawiro, S.	MJR	Photogeology of volcanic terrains with particular reference to a photo-interpetation of western Vanua Levu, Fiji.
	Le Duy, P.	KSWC	Late Permian Actinopterygian fishes from Blackwater, Queensland.
	Reid, E. J.	BWC	The geology and geochemistry of the Murrumbucka region.
	Schmidt, B. L.	L. Gustafsen (RSES)/ NW	Geology of the Elura Ag-Pb-Zn deposit, Cobar district, NSW.
1982	Hough, D.	KAWC	The influence of parent rock on the nature of surficial materials in the Eden region.
1984	Kavalieris, I.	JLW	The geology and geochemistry of the Gunang Pani gold prospect, north Sulawesi, Indonesia.
1985	Banfield, J.	KAWC	The mineralogy and chemistry of granite weathering.
	Holzhauer, C.	RAE	Weathering of Fe-Ti oxide minerals in basalts.
1986	Buskas, A.	KSWC	Taxonomy and significance of lower Devonian silicified fenestrates from Taemas, NSW.
1988	Magee, J. W.	KAWC/D. Lock	Chemical and clastic sediments and late Quaternary history, Prungle Lakes, NSW.
	Platts, W.	JLW	A study of the base-metal, gold and barite mineralisation in the volcano-sedimentary belts of the Canberra region. [2 vols]
1989	Hill, P.	KAWC	Geophysics, structure and evolution of selected South Pacific seamounts and plateaus.
1992	Aspandiar, M.	RAE	Weathering of a basalt microsystems and mineralogy.
	Rumble, C.	RAE	Multi-element geochemistry and lead-isotope analysis of a mineralised granite and rhyolite: northern Qld applications to exploration in the regolith.
1993	Deacon, G.	KAWC	Shoreline sedimentation at modern plate boundaries—Huon Gulf, PNG.
	Marks, A.	RAE	Remote sensing of the regolith, Shoalwater Bay area, Qld.
	Swanson, K.	PDD	Late Quaternary and recent benthic ostracoda from the eastern Tasman Sea.
	Hancock, G.	RAE	The effect of salinity on the concentrations of radium and thorium in sediments.
1994	Carson, L.	MJR	Tectonic evolution of the Cullarin and Canberra blocks, near Queanbeyan, NSW.

Appendix 4

Year	Student	Supervisor	Title
	Crowe, W.	DJE	Geology, metamorphism and petrogenesis of the Fisher Terrane, Prince Charles Mountains, east Antarctica.
1995	Young, D.	DJE	Petrology of the Mawson charnockites, Antarctica.
2000	Kovacs, N.	P. Blevin	Magmatic and hydrothermal evolution of the Browns Creek Intrusive Complex and associated gold mineralisation.
2002	Qopoto, C.	RJA	Comparative geochemistry of some volcanic suites of Solomon Island and Bouganville: applications for metallogenesis.
2004	Patia, H.	RJA	Petrology and geochemistry of the recent eruption history at Rabaul Caldera, Papua New Guinea: implications for magmatic processes and recurring volcanic activity.

Appendix 5

List of PhD Theses

Year	Student	Supervisor	Title
1964	Jensen, A. R.	KAWC	The geology of the Blenheim area, Bowen Basin, Qld.
	Johnson, N. E. A.	DAB/KAWC	The geology of the Krawaree area, NSW.
	Wilson, E. G.	KAWC/MJR	The geology of the Farrar–Hoskinstown area, NSW.
1964	Belford, D. J.	DAB	The smaller foraminifera of the Miocene and Pliocene of Papua and New Guinea. (2 vols)
	Powell, N. A	DAB	The morphology, systematics, and distribution of some recent polyzoa (bryozoa) from New Zealand.
	Stauffer, M. R	KAWC/MJR	Multiple folding and deformation of Lower Paleozoic rocks near Queanbeyan, NSW.
1965	Abbott, M. J.	AJRW	A petrological study of the Nandewar volcano, NSW.
	Harris, J. F.	KLW	Metallogenic studies in south eastern New South Wales.
1966	Chappell, B. W.	AJRW	Petrogenesis of the granites at Moonbi, NSW.
	Steiner, J.	KAWC	Depositional environments of the Devonian rocks of the Eden-Merrimbula area, NSW. (3 vols)
1967	Kemezys, K. J.	KSWC	Studies in fossil brachiopod morphology.
1968	Day, R. W.	KSWC	Biostratigraphy and taxonomy of Lower Cretaceous molluscan faunas from the Queensland portion of the Great Artesian Basin.
	Gostin, V. A.	KAWC	Stratigraphy and sedimentology of the Lower Permian sequence in the Durras-Ulladulla area, Sydney Basin, NSW.
	Moss, A. J.	KAWC	Bed-load deposits of shallow, undirectional currents.
	Williams, K. L	JFL (RSES)	Hydrothermal zoning: a study of the lead-zinc ores of Zeehan, Tasmania.
1969	Bailey, J. C.	BWC	Geochemistry and petrogenesis of volcano-plutonic formations in the Georgetown Inlier, north Queensland.
	Chatterton, B. D. E.	KSWC	Some aspects of the palaeontology, palaeoecology and biostratigraphy of the limestones of the Murrumbidgee group at Taemas, near Yass, NSW.
	Rhodes, J. M.	AJRW	The geochemistry of a granite-gabbro association at Hartley, NSW.
1970	Both, R. A.	KLW/RAE	Minor element geochemistry of sulphide minerals in the Broken Hill lode, NSW.

Geology at ANU (1959-2009)

Year	Student	Supervisor	Title
	Jakeš, P.	AJRW	Analytical and experimental geochemistry of volcanic rocks from island arcs.
	Jones, B. G.	KAWC	Stratigraphy and sedimentology of the Upper Devonian Pertnjara and Finke groups, Amadeus Basin, NT.
	Joyce, A. S.	BWC	Geochemistry of the Murrumbidgee Batholith.
1971	Link, A. G.	KSWC	Stratigraphic developments of the Yass Basin.
1972	Bradley, G. M.	AJRW/R. Taylor (RSES)	The geochemistry of a medium pressure granulite terrain at southern Eyre Peninsula, Australia.
	Crain, I. K.	MJR	Statistical approach to the analysis of geotectonic elements.
	Haynes, D. W.	AJRW	Geochemistry of altered basalts (continental tholeiites) and associated copper deposits.
	Jell, P. A.	KSWC	Studies on Middle Cambrian trilobites: distribution and taxonomy.
	Jensen, A .R.	KAWC/CEBC	Permo-Triassic stratigraphy and sedimentation in the Bowen Basin, Queensland.
	Kesson, S. E.	BWC	Basic alkaline rocks.
1973	Chappell, J. M. A.	KAWC	Geology of coral terraces on Huon Peninsula, New Guinea.
	Reinson, G. E.	KAWC	Aspects of weathering and sedimentation in the Genoa River Basin and estuary, NSW - Victoria.
1974	Marjoribanks, R. W.	MJR	The structural and metamorphic geology of the Ormiston region, central Australia.
1975	Landrum, R .S.	KSWC	Biostratigraphy, palaeontology and palaeoecology of some Silurian and Devonian deposits in the Cobar Basin, NSW.
	Mason, D. R.	JAMcD	Geochemistry of intrusive rock suites and related porphyry copper mineralization in the Papua New Guinea-Solomon Islands region.
	Owen, J .A.	KSWC	Palynology of some Tertiary deposits from New South Wales.
1976	Hough, M .J.	BWC/JAMcD	Archean ultramafic metavolcanics host to nickel-sulphide mineralisation Mt Edwards, WA.
	Smith, I. E.	AJRW	Volcanic rocks from south-east Papua.
	Taylor, G. M.	KAWC	Barwon River NSW — Basin fill by low-gradient streams.
1977	McKirdy, D. M.	ECBC	Diagenesis of microbiol organic matter: a geochemical classification and its use in evaluating the hydrocarbon-generating potential of Proterozoic and Lower Palaeozoic sediments, Amadeus Basin, central Australia.

Appendix 5

Year	Student	Supervisor	Title
	Wyborn, L. A .I.	BWC	Aspects of the geology of the Snowy Mountains region and their implications for the tectonic evolution of the Lachlan Fold Belt.
1978	Willink, R .J.	KSWC	Permian Crinoids from Eastern Australia.
1980	Hayden, P.	MJR	Structure and metamorphism of the Ordovician rocks of the Cullarin Horst near Jerangle, NSW.
	Whalen, J. B.	JMcD	Aspects of granites and associated mineralization.
1981	Bhatia, M. R.	KAWC	Petrology, geochemistry and tectonic setting of some flysch deposits.
	Britten, R. M.	JMcD	The geology of the Frieda River Copper prospect, Papua New Guinea.
1983	Alam, M.	KAWC/GMT	Sedimentology and stratigraphic implications of a major inland distributive fluvial system; the Castlereagh drainage basin, NSW.
	Wyborn, D.	BWC	Fractionation processes in the Boggy Plain zoned pluton.
	Windrim, D.	WEC	Chemical and thermal evolution of Strangways granulites, central Australia.
1984	Feary, D. A.	KAWC	Facies, environments and evolution of a Silurian siliciclastic carbonate sequence, Boambolo, NSW.
1985	Johnston, .P	KSWC	Morphology, Relationships and Palaeoecology of Lower Devonian Bivalves from southeastern Australia.
	Sawka, W.	BWC	Geochemistry of differentiation process in granite magma chambers.
1987	Halley, S.	JLW	Genesis of the Mount Bischoff tin deposit.
	Hergt, J.	BWC	The origin and evolution of the Tasmanian dolerites. (2 vols)
1988	Kiene, W. E.	KAWC/ KSWC	Biological destruction on the Great Barrier Reef.
	Loosveld, R.	MJR/ M. Etheridge	Structure and techno-thermal history of the Eastern Mount Isa Inlier, Australia.
	Wang Qi-ming	RAE	Mineralogical aspects of Monzonite alteration: investigation by electron microscopy and chemistry.
	Wenlong Zang	KSWC	Analysis of Late Proterozoic-Early Cambrian microfossils and biostratigraphy in China and Australia. (3 vols)
1989	McKay, W. J.	JLW	A study of the geological setting, nature and genesis of the Woodlawn base metal deposit, NSW Australia.
	Shuang K. Ren	JLW	The Ardlethan tin field, NSW; breccia pipes and mineralization.
1990	De Caritat, P.	JCT	Aspects of sediment diagenesis: Empirical investigation (Denison Trough, Qld) and theoretical modelling.

Geology at ANU (1959–2009)

Year	Student	Supervisor	Title
	Stuart-Smith, P.	MJR/ M.Etheridge (AGSO)	Structure and tectonics of the Tumut Region, Lachlan Fold Belt, SE Australia.
	Swift, M.	JCT	Heat flow studies of the Exmouth Plateau offshore NW Australia.
1991	Champion, D.	BWC	The felsic granites of far north Queensland, Vol 1 & 2.
1993	Correge, T.	PDD	Late Quaternary palaeoeanography of the Qld Trough (western Coral Sea) based on Ostracoda and the chemical composition of their shells.
	Liu Keyu	KAWC	Sedimentation and tectonics of the Markham suture. PNG.
	Martinez, I.	PDD	Late Pleistocene Palaeoceanography of the Tasman Sea.
	Skirrow, R. G.	JLW	The genesis of gold-copper-bismuth deposits, Tennant Creek, Northern Teritory.
	Pigram, C. J.	JCT	Carbonate platform growth demise and sea-level record; Marion Plateau, Northeastern Australia.
1994	Passlow, V.	PDD	Late Quaternary history of the Southern Ocean offshore southeastern Australia, based on deep-sea Ostracoda.
	Tilley, D.	RAE	Models of bauxitic pisolith genesis : data from Weipa, Queensland.
	Huq, N. E	JCT/G. Brierley/ J.Chappell	Clarence River Lowland floodplain: a morphostratigraphic analysis of a complex Holocene floodplain.
	Kilpatrick, J.	DJE	The petrogenesis of magmatic charnockites.
1996	Chi, Ma	RAE	The ultra-structure of kaolin.
	Davis, A.	D. Ride	Quaternary mammal faunas and their stratigraphy in the northern Monaro region southeastern Australia.
	Moore, C. L.	RAE	Processes of chemical weathering of selected Cainozoic Eastern Australian basalts.
	Liang, T. C.	JCT/R. Korsch (AGSO)	Sedimentary environments and sequence stratigraphy of non- marine Intercratonic deposits; lower to middle Jurassic of the Surat Basin, Australia.
	Ding, C.	JCT	An object-based approached for studying reservoir connectivity.
1997	Kjølle, I.	JLW	The setting and genesis of the Browns Creek gold-copper skarn deposit, NSW, Australia.
	Armand, L. K.	PDD	The use of diatom-transfer functions in estimating sea-surface temperature and sea-ice cores from the southern Indian Ocean.
1998	Aspandiar, M.	RAE	Regolith and Landscape Evolution of The Charters Towers Area, north Queensland.

Appendix 5

Year	Student	Supervisor	Title
	Cameron, G.	JLW	The Hydrothermal Evolution and Genesis of the Porgera Gold Deposit, Papua New Guinea.
	Sha, Lian-K.	BWC	Order-Disorder Kinetics of Atoms in Crystals, and Phosphorus Geochemistry of Granites, with Implication for Lunar Rocks and Meteorites.
	Sobhan, N.	JCT/MJR/B. Jones	Depositional Architecture and History of the Late Permian Broughton, Pheasants Nest and Erins Vale Formations, Southern Sydney Basin, NSW, Australia.
2000	Bryant, C.	RJA	Towards understanding the temporal geochemical evolution of intraoceanic arc systems : a case study of Izu-Bonin and Mariana Tephras.
	Troitzsch, U.	DJE	The crystal structure and thermodynamic properties of titanite solid-solution Ca (Ti,A1) (O,F) SiO.
	Radke, L.	PDD	Solute Divides and Chemical Facies in Southeastern Australian Salt Lakes and the Response of Ostracods in Time (Holocene) and Space.
	Hill, S.	RAE	The Regolith and Landscape Evolution of the Broken Hill Block, western NSW, Australia.
	Bastrakov, E.	JW	Gold metallogenesis at the Lake Cowal Prospect, NSW.
2002	Tomkins, A.	DJE	Evolution of the granulite-hosted Challenger Gold Deposit, South Australia: Implications for ore genesis.
2003	Murgese, S.D.	PDD	Late Quaternary palaeoceanography of the eastern Indian Ocean based on benthic foraminifera.
2004	Bierwirth, P.	PNC	Methods of spectral geology utilizing airborne-hyperspectral and satellite remote sensing-applications for exploration and acid-mine drainage.
	McPherson, A.	RAE	Salt Sources and Development of the Regolith Store in the Upper Billabong Creek Catchment, southeast NSW.
	Spandler, C.	JM/RJA	The geochemical and petrological evolution of subduction zones: Insights from blueschist to eclogite-facies rocks from New Caledonia and high pressure hydrothermal experiments.
	Bostock, H.	BNO	Geochemically tracing the intermediate and surface waters in the Tasman Sea, southwest Pacific.
	Heithersay, P. S.	JLW	The shoshonite-associated Endeavour 26 North porphyry Cu-Au deposit, Goonumbla NSW.
2005	Holgate, F.	PNC	Exploration and Evaluation of the Australian Geothermal Resource.
	Young, M.	PDD	The distribution of organic-and calcareous-walled dinoflagellate cysts from the easternIndian Ocean;a proxy for late quaternary palaeo-oceanographic reconstructions.
2006	Budd, A.	DJE	The Tarcoola Goldfield of the central Gawler Gold Province, and the Hitabo Association granites, Gawler Craton South Australia.

Geology at ANU (1959–2009)

Back piece End-of-year field excursion to Chile, January 2008: Jorisques Volcano with Licancabur Volcano behind on the left. Front row (kneeling) from left to right: Kate Boston, Sally Mayberry, Eva Reynolds, Tarum Whan, Frances Jenner, Tim Curran. Back (standing) from left to right: Alex Johnston, Tim Richmond, Iona Stenhouse, Sarlae McAlpine, Tristan Webber, Sarah O'Callaghan, Louise Soroka, James Hughes, Simon Richmond, Dominic Greenslade, Clare Firth, Jane Thorne, Max Collett, Moira Goddard, Rhiannon Mann, Margaret May, Jennifer Burke, Scott Biddlecombe, Helen Tait, 'Ruben' (bus driver) and Richard Arculus.

www.ingramcontent.com/pod-product-compliance
Lightning Source LLC
Chambersburg PA
CBHW061938290426
44113CB00029B/2916